FOOD CONTAMINANTS
Sources and Surveillance

Food Contaminants
Sources and Surveillance

Edited by

Colin Creaser
University of East Anglia, Norwich

and

Rupert Purchase
British Industrial Biological Research Association, Carshalton, Surrey

THE ROYAL SOCIETY OF CHEMISTRY
1841-1991

Learning Resources
Centre

ISBN 0-85186-606-9

A catalogue record for this book is available from the British Library

Published by The Royal Society of Chemistry.
Thomas Graham House, Science Park, Cambridge
CB4 4WF

Typeset by Keytec Typesetting Ltd., Bridport, Dorset, UK
Printed by Redwood Press Ltd, Melksham, Wiltshire

Preface

This book contains contributions based on the proceedings of two symposia on food contamination held in London in April 1989 and May 1990, both of which were organized jointly by the Environment, Food Chemistry, and Toxicology Groups of the Royal Society of Chemistry.

The aim of these meetings was to assess the extraneous chemical contamination of food from two sources: firstly, *food-chain contaminants* – the presence of plant toxicants or fungal metabolites in food, or the contamination of food from environmental sources (airborne, aquatic, and terrestrial); secondly, *food-production contaminants* – contaminants of man-made origin brought about by a desire to facilitate food production and distribution.

The surveillance of food contaminants through analytical and toxicological investigations has stimulated public awareness, and in turn has led to changes in the control of these substances by the legislators. Important as these issues are, they should be placed in the broader context of food hazards in general. A list of hazards associated with food drawn up by the FDA in the United States, and reiterated by the late Professor Leon Golberg, a pioneer of toxicology both in the UK and the USA (*Chem. Ind. (London)*, 1982, 354), is as follows (in decreasing order of importance): food-borne disease of microbial origin, malnutrition, environmental contaminants, toxic normal constituents of food, pesticide residues, and food additives. It is through the vigilance of the agriculture and food industries and legislative bodies that contamination problems of all these types are comparatively rare.

The contributions in this volume concentrate on the contamination of food by chemicals arising from environmental and food-production sources. Chapter 1 is concerned with food-chain contaminants present in food as natural components of the diet. This is followed by discussion of the chlorinated dioxins and furans (Chapter 2), and polycyclic aromatic hydrocarbons (Chapter 3). After an introduction to the control and surveillance of food-production contaminants (Chapter 4), four areas of activity are described: migration from food contact materials with particular reference to plastics (Chapters 5 and 6), the regulatory control and analysis of veterinary products (Chapters 7 and 8 respectively), the analysis of pesticides in drinking water (Chapter 9), and finally the problem of food taints (Chapter 10).

We thank our colleagues from the RSC subject groups for their contribution to the organization of the two symposia which led to the publication of this book: Dr. David Henshall, Dr. Robert Massey, Dr. Martin Shepherd, and Dr. David Taylor.

Colin Creaser and Rupert Purchase
March 1991

Contents

Chapter 1 Natural Toxicants in Food
Jenny A. Lewis and G. Roger Fenwick   1

Chapter 2 Polychlorinated Dibenzo-*p*-dioxins, Polychlorinated
Dibenzofurans, and the Food Chain
James R. Startin   21

Chapter 3 Analysis and Occurrence of Polycyclic Aromatic
Hydrocarbons in Food
Keith D. Bartle   41

Chapter 4 Food Production Contaminants: Control and Surveillance
Jonathan R. Bell and David H. Watson   61

Chapter 5 Toxicology and Regulatory Control of Components of
Food Contact Plastics
Rupert Purchase   73

Chapter 6 Contaminants from Food Contact Materials:
Analytical Aspects
John Gilbert   85

Chapter 7 Use and Regulatory Control of Veterinary Drugs in
Food Production
Kevin N. Woodward   99

Chapter 8 Analysis of Veterinary Drug Residues in Edible
Animal Products
Martin J. Shepherd   109

Chapter 9 Analysis of Pesticides at Low Levels in Drinking Water
Keith M. Moore   177

Chapter 10 Unwanted Flavours in Food
J. David Henshall   191

Subject Index   201

Contributors

Keith D. Bartle, *School of Chemistry, University of Leeds, Leeds LS2 9JT, UK*

Jonathan R. Bell, *Ministry of Agriculture, Fisheries and Food, 17 Smith Square, London SW1P 3JR, UK*

G. Roger Fenwick, *AFRC Institute of Food Research, Colney Lane, Norwich NR4 7UA, UK*

John Gilbert, *Ministry of Agriculture, Fisheries and Food, Food Science Laboratory, Colney Lane, Norwich NR4 7UQ, UK*

J. David Henshall, *Campden Food and Drink Research Association, Chipping Campden, Gloucestershire GL55 6LD, UK*

Jenny A. Lewis, *AFRC Institute of Food Research, Colney Lane, Norwich NR4 7UA, UK*

Keith M. Moore, *Water Research Centre plc, Henley Road, Medmenham, Marlow, Buckinghamshire SL7 2HD, UK*

Rupert Purchase, *The British Industrial Biological Research Association, Woodmansterne Road, Carshalton, Surrey SM5 4DS, UK*

Martin J. Shepherd, *Ministry of Agriculture, Fisheries and Food, Food Science Laboratory, Colney Lane, Norwich NR4 7UQ, UK*

James R. Startin, *Ministry of Agriculture, Fisheries and Food, Food Science Laboratory, Colney Lane, Norwich NR4 7UQ, UK*

David H. Watson, *Ministry of Agriculture, Fisheries and Food, 17 Smith Square, London SW1P 3JR, UK*

Kevin N. Woodward, *Veterinary Medicines Directorate, Woodham Lane, New Haw, Weybridge, Surrey KT15 3NB, UK*

CHAPTER 1

Natural Toxicants in Food

JENNY A. LEWIS AND G. ROGER FENWICK

1 Introduction

The maintenance of the safety of the food supply is an essential function of the state. The food industry has a statutory obligation to provide safe, wholesome food for its customers but it is the duty of government to ensure that satisfactory methods and understanding exist to fulfil this obligation. As has become all too apparent in the UK recently, when agricultural products cannot be sold by virtue of the presence of recognized hazards, producers lose income, trade is damaged, and consumers are justifiably concerned. Public awareness of the importance of safe, highly nutritious food has never been higher, and consumers are actively seeking factual information and reassurance about the safety of their diets and of the individual foods they consume. Such concern and interest is to be welcomed, even at the expense of the sometimes ill-informed, and frequently emotional, arguments which can develop as a consequence.

Much public concern has been focused on chemicals in the food chain, be these agrochemicals, post-harvest dips or sprays, recognized and approved additives, environmental contaminants, or adulterants. There has, in comparison, been much less attention paid to potentially harmful chemicals which occur naturally in our diet as a consequence of their presence in plants used as foods and animal feeding stuffs. Such compounds, usually termed natural toxicants, are the subject of this chapter. Serious health problems due to natural toxicants are not new; indeed one such episode is described in the Book of Numbers, Ch. 11, vv. 31–33. Another, admittedly extreme, example from history was the periodic flare-up of ergotism. This occurred for over a thousand years and into the present century, claiming many hundreds of thousands of victims. It is now recognized that the causes of the epidemics were the naturally occurring alkaloids produced by the rye fungus, *Claviceps purpurea*, and it is interesting to note that the diminution of the problem was not a consequence of any public or domestic health measures, but rather a change in social and agricultural practice with the potato replacing rye as a staple of the European diet.

1

In general, however, the effects of natural toxicants are more likely to be chronic rather than acute and consequently, even when detected and measured, they are frequently impossible to relate to individual dietary components, let alone particular chemical constituents thereof. Food is probably the most chemically complex material normally encountered by the population at large. The number of naturally occurring compounds in food plants probably exceeds half a million, ranging from volatile flavour compounds to macromolecular proteins or polysaccharides. The preparation of food, involving a variety of processes from mixing to chopping, boiling, and frying, produces complex secondary chemical reactions, the nature and extent of which are only now becoming clear to the food chemist. Whilst many of these reactions are perceived by the senses as desirable, *e.g.* in the development of anticipated flavour, texture, and colour, this is not necessarily the case.

Despite the continuing development and refinement of techniques for the separation, isolation, and structural determination of individual chemical compounds, only a minority of those present in food have been identified and very many fewer have been subject to biological scrutiny at anything like the level carried out on agrochemicals, pollutants, food additives, and colorants. Various reasons may be suggested for this state of affairs, including the cost of pharmacological and toxicological testing, the frequent difficulties encountered in the isolation of these natural food components in amounts and purities sufficient for biological examination, and, perhaps most importantly, the overemphasis on the screening of those exogenous substances in food which are generally believed by the public to be the only 'chemicals' present and which have become a major focus of attention for the media and pressure groups. It is unfortunate that food scientists, and food chemists in particular, have so singularly failed to correct this misapprehension and to explain to the public that, in the context of food, the word *chemical* refers to both natural and man-made compounds and that, in contrast to what the advertizing companies would lead us to believe, 'natural' is not necessarily synonymous with 'wholesomeness' and 'safety'.

2 The Risk Due to Natural Toxicants

The six principal categories of food hazard are listed in rank order in Table 1. This ranking, based upon objective scientific criteria (including the severity, incidence, and onset of biological symptoms) was determined by Wodicka in 1971.[1] Around the same time Hall (1971)[2] pointed out that the ranking was not linear and that the hazard associated with environmental pollutants and natural toxicants was $\sim10^3$-fold less than that originating from nutrient imbalance (which embraces both excess and deficiency). Furthermore, contrary to media comment and public concern, the risk due

[1] V. O. Wodicka, *Food Chem. News*, 1971, March 1st, 130.
[2] R. L. Hall, *Flavour Ind.*, 1971, August, 455.

Table 1 *Priority of food hazards (Wo-dicka, 1971)*

1 Microbial contaminants
2 Nutritional imbalance
3 Environmental contaminants
4 Natural toxicants
5 Pesticide residues
6 Food additives

to the presence of pesticide residues or food additives is considered to be $\sim 10^2$-fold lower than that resulting from natural toxicants.

In 1971, Hall[2] described subjective rankings obtained as a result of the comments of the press, food industry, 'fringe hysteria' (presumably the opposite of informed public opinion), and the US FDA regulatory authorities. The only common feature was that natural toxicants were ranked least important by all these groups. A recently constructed questionnaire, used by staff at the Institute of Food Research in 1988–89 (R. Shepherd and G. R. Fenwick, unpublished) confirms this finding and supports the comment of the 1985 UK Food Safety Research Consultative Committee that 'there is an almost complete mismatch between areas of known quantifiable hazard on the one hand and public concern and legislative activity on the other'.

One reason for this situation has been mentioned above, the general absence of readily identified, and acute, symptoms associated with natural toxicants. The terminology used may also be confusing; if natural, in the public mind, is indicative of goodness, freshness, and wholesomeness and encapsulates an ecologically desirable 'chemical-free', simple state of living, then toxicant is perceived as being related to food poisoning or linked to pollution, adulteration, and contamination of the food supply. Thus the term 'natural toxicant' offers an apparent contradiction, describing an area of hazard little understood or regarded by consumers and legislators alike.

Paradoxically, as inspection of any of a number of books on natural toxicants will confirm, there is no shortage of information available on the biological effects of chemicals isolated from plants, including those consumed by animals and man (for example Cheeke, 1989[3]). The problem is that the majority of such studies have used *in vitro* or other assays, the relevance of which to the human condition is open to question. There is no doubt that the techniques available to the chemist for analysing food components have generally far outstripped the complementary biological information which is necessary to put the analytical data into a scientific and social context. This in turn, makes assessment of research priorities difficult for the administrator. In the next sections the elements of a coherent natural toxicant research strategy are discussed and criteria for assessing priorities are suggested.

[3] 'Toxicants of Plant Origin', ed. P. R. Cheeke, Volumes I–IV, CRC Press Inc., Boca Raton, Florida, 1989.

3 Natural Toxicant Research

As part of an overall food safety policy, a research programme on natural toxicants should have the elements indicated in Figure 1. Following the identification and characterization of an effect, which might be acute or chronic, the causative principle(s) should be separated and identified. Methods of analysis should then be developed to enable calculation of exposure data on both the population at large and sub-groups judged to be particularly at-risk. It may also be necessary to develop additional methods of analysis to facilitate the clinical or toxicological studies which are also an integral part of this programme. It will be necessary to isolate sufficient of the active principle(s) to optimize and validate these analytical studies and also to facilitate the biological investigations. The aim of the latter should be to determine the potency of the identified compounds in man or in suitable animal models judged to have relevance to the human condition. Ideally toxicological investigations should include examination of the mechanism underlying the observed effect since this may enable extrapolation to chemical compounds having structural similarities. The study of selected compounds under conditions of long-term, low-level exposure (although, as will be seen later, exposures are not necessarily low) would, arguably, be of more benefit than wider screenings using tissue culture or micro-organisms.

Natural Toxicant Research

Figure 1 *The basic elements of a coherent programme on natural toxicants*

Any risk associated with consumption of a particular foodstuff containing identified natural toxicants will be determined by the exposure data, the metabolic activation or detoxification of the active principle(s), and the biological potency. Should such calculation result in the identification of a risk then the question of reducing or removing the risk has to be considered. Options for this will be considered below.

It may be predicted, with some confidence, that exposure to a variety of natural toxicants will increase during the present decade. There are various reasons for this. First, as a result of nutritional advice and recommendations, consumption of green vegetables, plant protein, and fibre will increase. Although this advice is certainly correct it must be remembered that such dietary changes will be associated with increased exposure to a range of biologically active secondary metabolites whose long-term effects, for good or ill, on human health cannot yet be ascertained. Secondly, there are dietary fashions – for example, the consumption of potato peel referred to elsewhere. The increased reliance of some individuals on 'health' foods and herbal beverages and tonics may also be seen in this context. In most cases the botanical composition, and hence chemical content, is unknown. Exposures may currently be very large and continuous – very different from earlier times when the plants, their decoctions, and products were used infrequently against specific illnesses, or in response to particular symptoms. A third reason, amplified below, relates to changes in agricultural practice, with external (agrochemical) plant protection being replaced by 'natural' processes – often associated with the compounds considered here, in the context of food, as natural toxicants.

Finally it should be emphasized that attention should not be uncritically directed at toxic compounds in plants *per se*, because they may subsequently be rendered inactive by storage, processing, metabolism or digestion. Conversely inactive species may be activated by chemical transformations caused by the same processes.

4 Criteria for Assessing Research Priorities

Suggested criteria are listed in Table 2 and will be discusssed individually below.

Table 2 *Criteria for priority assessment*

Epidemiological relationships
Physiological/diet-related disturbances in man
Performance data on animals
Extent of use of wild types in plant breeding
Physicochemical/chemical effects in non-food systems
Serendipity

(i) Epidemiological relationships are the most difficult to identify, especially as diets become more and more internationalized. Examples generally relate to the dietary habits of particular groups of people in particular societies and include, for example, oesophageal and gastric cancers in Japan and South America related to bracken-derived carcinogens (Fenwick, 1988[4]), haemolytic syndromes associated with legume consumption in individuals of Mediterranean extraction (Chevion *et al.*, 1983[5]), and

[4] G. R. Fenwick, *J. Sci. Food Agric.*, 1988, **46**, 147.
[5] M. Chevion, J. Mager and G. Glaser, in 'Handbook of Naturally-occurring Food Toxicants', ed. M. Rechcigl, jr., CRC Press Inc, Boca Raton, Florida, 1983, pp. 63–79.

hypoglycaemia in Jamaica which has been shown to be due to consumption of the unripe fruit of the ackee tree (Kean, 1976[6]).

(ii) Diet-related disturbances in man, though infrequent, do enable studies to be focused on particular diets, and even foodstuffs, so that cause/effect relationships may often be readily identified. Thus, for example, the acute physiological disturbance which follows consumption of green or damaged potatoes (see below) or of squash or zucchini containing high levels of cucurbitacins (Rymal *et al.*, 1984[7]), severe clinical symptoms due to the consumption of certain crustacea and Pacific fish (Schantz, 1973[8]), and the problems of weakness which have been shown to relate to excessive consumption of liquorice confectionery containing the sweet saponin, glycyrrhizin (Lutomski, 1983[9]). It is not necessary for food plants to be consumed to cause problems; thus celery pickers are susceptible to severe dermatitis, the result of contact with photocarcinogenic psoralens, the accumulation of which in the plant may be increased as a consequence of damage, infection, or stress (Beier *et al.*, 1983[10]). Finally cassava (manioc) is a dietary staple in many parts of the world; nevertheless, if not properly processed before being consumed it may cause death or severe illness as a result of the presence of glycosides which yield hydrogen cyanide on hydrolysis. These compounds may also cause chronic effects including amblyopia, damage of the optic nerve, tropical atoxic neuropathy, and thyroid disorders (Conn, 1973[11]).

(iii) In some circumstances, performance data from animals may be useful, for example reproductive disturbances in farm and other animals grazing on pasture legumes or consuming soya-based rations which are known to contain isoflavone- and coumestan phytoestrogens (Price and Fenwick, 1985[12]), damage to liver and lungs in cattle fed sweet potatoes which is now known to be due to the presence of furanoterpenes (Wilson *et al.*, 1978[13]), and severe illness and death in animals exposed to a range of pyrrolizidine alkaloids occurring in pasture weeds (World Heath Organization, 1988[14]). In all these cases the same compounds may be regularly consumed by humans.

[6] E. A. Kean, 'Hypoglycin', Academic Press, New York, 1976.

[7] K. S. Rymal, O. L. Chambliss, M. D. Bond, and D. A. Smith, *J. Food Protect.*, 1984, **47**, 270.

[8] E. J. Schantz, in 'Toxicants Occurring Naturally in Foods', ed. F. M. Strong, National Academy of Sciences, Washington, DC, 1973, pp. 424–447.

[9] J. Lutomski, *Pharm. Unserer Zeit*, 1983, **12**, 49.

[10] R. C. Beier, G. W. Ivie, and E. H. Oertli, in 'Xenobiotics in Foods and Feeds', ed. J. W. Finley and D. E. Schwass, American Chemical Society, Washington, 1983, pp. 295–310.

[11] E. E. Conn, in 'Toxicants Occurring Naturally in Foods', ed. F. M. Strong, National Academy of Sciences, Washington, 1973, pp. 299–308.

[12] K. R. Price, and G. R. Fenwick, *J. Food Add. Contam.*, 1985, **2**, 73.

[13] B. J. Wilson, J. E. Garst, R. D. Linnabary, and A. R. Doster, in 'Effects of Poisonous Plants on Livestock', ed. R. F. Keeler, K. R. Van Kampen, and L. F. James, Academic Press, New York, 1978, pp. 311–333.

[14] World Health Organization, 'Pyrrolizidine Alkaloids, Environmental Health Criteria 80', World Health Organization, Geneva, 1988.

However, care has to be taken not to extrapolate uncritically from the results of animal, or *in vitro*, experiments to man. To cite but one example, severe haemolytic anaemia, possibly with fatal consequences, may result from the ingestion of cruciferous forages by cattle and other animals. The haemolytic precursor has been identified (Smith *et al.*, 1974[15]) as *S*-methylcysteine sulphoxide (SMCO), a compound also found in brassica vegetables, beans, and alliums. However, these foods pose no risk to human health since in man, unlike the above animals, SMCO is not broken down to the active haemolysin, dimethyl disulphide. Glucosinolates (formerly termed mustard oil glycosides) are present in animal feeding stuffs such as oil seed rape, and considerable effort and expenditure have been directed at reducing the exposure of farm animals to these compounds, or more correctly to the various products of their chemical/enzymic breakdown. Possible consequences of the presence of such compounds in the human diet will be considered later.

(iv) The effects of the use of wild-type plants (*i.e.* those which are not cultivated and frequently inedible) in breeding programmes to achieve primary objectives, such as increased yield or disease resistance, have not been widely studied. Since increased disease resistance is in many instances due to increased levels of secondary metabolites which exhibit wide ranging biological activities, there is reason to be vigilant in this respect. A well documented, but highly relevant, example of the unexpected and potentially hazardous consequences of plant breeding is that of the potato. The cultivar Lenapé, possessing excellent processing and agronomic characteristics, was being introduced into the USA two decades ago after having completed all the then-applicable screens and trials. The sudden realization that very high levels of glycoalkaloids possessing anticholinesterase-activity could be produced under particular combinations of day-length and temperature in the North Eastern states of the USA led to the rapid removal of this variety (Zitnak and Johnson, 1970[16]) and a public health problem of major proportions was narrowly averted.

Investigation subsequently showed the problem to be a consequence of the use in the breeding programme of the wild potato *Solanum chacoense*. A rather similar situation, not well publicized, later arose in the UK and involved the use of another wild potato, *S. vernii*, which was being used to introduce eel-worm resistance. Very late on in the breeding programme material was screened for glycoalkaloid content and, as a result of disturbingly high levels, many potentially commercially attractive clones had to be withdrawn.

(v) Saponins are found in beans and peas, and their effectiveness in reducing plasma cholesterol levels in animals and man (see Price *et al.* 1987[17]) has led to suggestions that saponin exposure should be increased as

[15] R. H. Smith, C. R. Earl, and N. A. Matheson, *Trans. Biochem. Soc.*, 1974, **2**, 101.
[16] A. Zitnak and G. R. Johnson, *Am. Potato J.*, 1970, **47**, 256.
[17] K. R. Price, I. T. Johnson, and G. R. Fenwick, *CRC Crit. Rev. Food Sci. Nutr.*, 1987, **26**, 27.

a means of naturally reducing the risk of coronary artery disease; such increased exposure might be achieved by increased consumption of legumes, developing new varieties with elevated saponin levels, or *via* regular intakes of saponin-containing health supplements. However, such suggestions have to be treated with caution, since it is well known from a variety of uses that saponins possess potent surface-active properties. Studies have demonstrated that saponins may exhibit a significant effect on the gut wall, thereby increasing its permeability to other dietary components or xenobiotics (Johnson *et al.*, 1986[18]; Gee *et al.*, 1989[19]). These effects are strongly structure-dependent, suggesting in turn that analytical information on levels and exposure of individual saponins is more important than data based upon 'total saponin' intake.

(vi) The final criterion, serendipity, rarely receives due credit; in this regard natural toxicant research is no different from many other branches of scientific investigation. A good example is the finding of substantially increased levels of oestrogenic activity in human urine following microbial conversion of natural isoflavone glycosides (such as those in soya) to more active metabolites (Axelson *et al.*, 1984[20]). This observation stemmed from the routine screening of patients rather than from any overt examination of oestrogenic effects.

5 Natural Toxicants

5.1 Glycoalkaloids

Potatoes and other members of the Solanaceae contain toxic glycoalkaloids, mainly α-chaconine and α-solanine (Figure 2), which interfere with the normal function of the central nervous system and also irritate the gastro-intestinal tract. These compounds are present in highest quantities in the above ground parts (flowers, $3-5 \times 10^3$ p.p.m.; leaves 400–1000 p.p.m.) and in the sprouts ($2-4 \times 10^3$ p.p.m.). Levels in the tuber vary, and are found mainly in the skin and peel region (150–600 p.p.m.); the significance of this from the food preparation and processing standpoint is that peeling reduces the glycoalkaloid content to 10–50 p.p.m.

A variety of factors, including genetic origin, adverse post-harvest handling and storage, environment and location, physiological stress, and processing lead to the increased synthesis, and accumulation, of glycoalkaloids in the tuber (Jadhav *et al.*, 1981[21]). Acute illness, and even death, in animals and man have followed the consumption of damaged, rotten,

[18] I. T. Johnson, J. M. Gee, K. Price, C. L. Curl, and G. R. Fenwick, *J. Nutr.*, 1986, **116**, 2270.

[19] J. M. Gee, K. R. Price, C. L. Ridout, I. T. Johnson, and G. R. Fenwick, *Toxicol. In Vitro*, 1989, **3**, 85.

[20] M. Axelson, J. Sjövall, B. E. Gustafsson, and K. D. R. Setchell, *J. Endocrinol.*, 1984, **102**, 49.

[21] S. J. Jadhav, R. P. Sharma, and D. K. Salunkhe, *CRC Crit. Rev. Toxicol.*, 1981, **11**, 21.

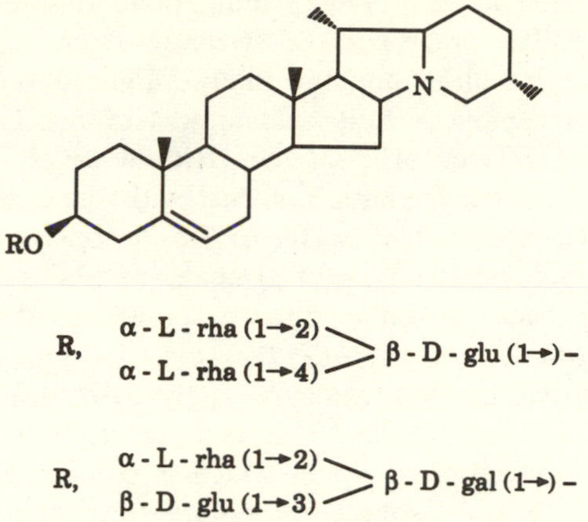

$$R, \quad \begin{matrix} \alpha\text{-}L\text{-}rha\,(1\to2) \\ \alpha\text{-}L\text{-}rha\,(1\to4) \end{matrix} \Big\rangle \beta\text{-}D\text{-}glu\,(1\to)\text{-}$$

$$R, \quad \begin{matrix} \alpha\text{-}L\text{-}rha\,(1\to2) \\ \beta\text{-}D\text{-}glu\,(1\to3) \end{matrix} \Big\rangle \beta\text{-}D\text{-}gal\,(1\to)\text{-}$$

Figure 2 *Structures of the potato glycoalkaloids, α-chaconine (upper) and α-solanine (lower)*

green, sprouted, or blighted potatoes – *i.e.* low quality, sub-standard produce. Both α-chaconine and α-solanine are potent cholinesterase inhibitors and symptoms associated with such pharmacological activity include drowsiness, difficult or laboured breathing, weakness, paralysis, and lack of consciousness. Their effects on the gastrointestinal tract lead to inflammation of the intestinal mucosa, haemorrhage or ulceration, abdominal pain, and diarrhoea.

The most recent outbreak of potato poisoning in the UK occurred ten years ago in Lewisham. Seventy-eight schoolboys were taken ill 7–19 hours after a meal (containing potatoes which were later found to contain levels of glycoalkaloids in excess of 300 p.p.m.); seventeen children required hospital treatment with three being described as seriously ill. Symptoms included sickness, headache, vomiting, diarrhoea, and abdominal pain (McMillan and Thompson, 1979[22]).

As a result of problems with Lenapé (see above), and other varieties, developed via the use of glycoalkaloid-rich wild type *Solanum* species, new methodologies were developed for the analysis of these compounds (see Coxon, 1984[23]). These proved inappropriate for the large-scale screening of breeding programmes and a simple, much more rapid, ELISA procedure was introduced (Morgan *et al.*, 1983[24]). Unlike food additives, there is no statutory control of glycoalkaloid levels in potatoes, but a generally recognized limit of 200 p.p.m. has been set – this being derived

[22] M. McMillan and J. G. Thompson, *Q. J. Med.*, 1979, **48**, 227.
[23] D. T. Coxon, *Am. Potato J.*, 1984, **61**, 169.
[24] M. R. A. Morgan, R. McNerney, J. A. Matthew, D. T. Coxon, and H. W.-S. Chan, *J. Sci. Food Agric.*, 1983, **34**, 593.

from the finding (Sinden *et al.*, 1976[25]) that above this level the potato becomes increasingly bitter and toxic. Most tubers from table varieties of potato are well below this limit, and the likely human exposure is further reduced if the peel is removed; thus calculations of the UK mean daily intakes have produced figures of 5–31 mg (Ridout *et al.*, 1988[26]). It is worth while emphasizing the dramatic contrast with the approach to food additives in processed food. If it were to be seriously suggested that compounds with the biological activity of glycoalkaloids should be added to any food, let alone such a staple one as potatoes, there would be considerable, and justified, concern.

Amongst certain individuals the exposure to glycoalkaloids may be much larger, perhaps four-fold or greater. This is a result of a current trend towards the consumption of potato peel which is considered, mistakenly, by some people to be a good source of dietary fibre and vitamins. The effects, *chronic* rather than acute, of regular consumption of the peel (from jacket potatoes, snack products made wholly from peel, or potato skins) are largely unknown; however recent studies (Gee *et al.*, 1989[19]) have demonstrated glycoalkaloids to possess potent gut-permeabilizing activity, and further work is in progress to examine the possible clinical consequences which may ensue if this results in the passage across the gut of other dietary components, for example those able to provoke the immune system.

5.2 Glucosinolates

Consumption of leafy and root brassica vegetables in the UK is greater than in other West European countries and North America. There has been a trend in recent years away from cabbage and Brussels sprouts, towards the more delicately-flavoured cauliflower, calabrese, and Chinese cabbage. All these vegetables contain glucosinolates, a complex class of sulphur-containing glycosides. Under the influence of an enzyme, myrosinase, also found in the plant, glucosinolates are hydrolysed to yield an unstable aglucone which subsequently decomposes to provide a range of products (dependent on R, Figure 3) The chemical nature and properties and the biological effects and potencies of these compounds are determined by the nature and amount of the parent glucosinolates and by the conditions of the breakdown (Heaney and Fenwick, 1988[27]).

A great deal of human and financial resource has been expended in many countries over the past two decades with the aim of reducing the intakes of glucosinolates in farm animals, the intake being a consequence of the use of rapeseed products in livestock rations. As a result new

[25] S. L. Sinden, K. L. Deahl, and B. B. Aulenbach, *J. Food Sci.*, 1976, **41**, 520.

[26] C. L. Ridout, S. G. Wharf, K. R. Price, I. T. Johnson, and G. R. Fenwick, *Food Sci. Nutr.*, 1988, **42F**, 111.

[27] R. K. Heaney and G. R. Fenwick, in 'Natural Toxicants in Food, Progress and Prospects', ed. D. H. Watson, Ellis Horwood, Chichester, 1988, pp. 76–109.

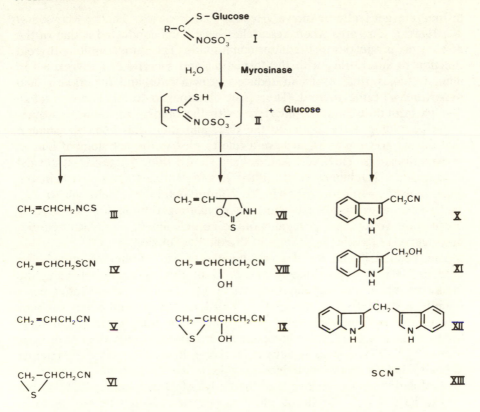

Figure 3 *Products of myrosinase-induced hydrolysis of glucosinolates*

'low-glucosinolate' varieties have been developed and now dominate the market place. The European Commission has assisted in this process by offering a premium to growers of such 'improved' varieties and has expressed the intention to remove all financial support from varieties having a higher glucosinolate content from 1991. The main problems associated with the presence of glucosinolates in animal rations are poor growth and performance, liver damage, impairment of normal thyroid function, and interference with normal metabolic processes. Such problems have been described in detail by Mawson *et al.* (1989).[28]

In comparison, considerably less attention has been devoted to the effect of glucosinolates and their products in man. Thus although many studies in animals and *in vitro* systems have shown isothiocyanates (III, Figure 3) and 5-vinyloxazolidine-2-thione (VII) to inhibit normal enzymatic processes, nitriles (VIII) to be hepatotoxic, epithionitriles (VI, IX) to possess weak mutagenic activity, and both 5-vinyloxazolidine-2-thione and thiocyanate ion (XIII) to interfere with thyroid function, the effects of such compounds

[28] R. Mawson, R. K. Heaney and G. R. Fenwick, in 'Natural Toxicants – Volume II, Glycosides', ed. P. R. Cheeke, CRC Press Inc., Boca Raton, Florida, 1989, pp. 1–41.

in man are generally unknown (Heaney and Fenwick, 1988[27]; Mawson *et al*., 1989[28]). One area where researches have been conducted is that of the goitrogenic properties of the latter compounds. The thione inhibits thyroid function by interfering with the formation and secretion of thyroxine. In animal studies this results in reduced growth rate and hyperplasia and hypertrophy of the thyroid. The effect of thiocyanate ion (which, unlike 5-vinyloxazolidine-2-thione, interferes with the binding of iodine), is overcome by adding extra iodine to the diet, and there has been speculation that dietary goitrogens might have a role to play in the aetiology of human thyroid disorders. However, studies have shown thiocyanate ion at normal dietary levels (Dahlberg *et al*., 1984[29]), and Brussels sprouts rich in the glucosinolate precursor of 5-vinyloxazolidine-2-thione (McMillan *et al*., 1986[30]) to have no adverse effect on circulating thyroid hormones. It is possible, however, that in regions of iodine deficiency, a brassica-rich diet may aggravate existing symptoms of thyroid dysfunction.

The mean daily intake of glucosinolates of 5% of the UK population exceeds 300 mg, approximately one third of this figure representing the intake of indole glucosinolates (Sones *et al*., 1984[31]). These latter compounds, particularly rich in varieties of Brussel sprouts and Savoy cabbage (see Heaney and Fenwick, 1988[27]), have attracted attention in recent years for two reasons. First, these compounds, or more correctly their hydrolysis products (X–XII, Figure 3) have been shown in studies with experimental animals, and in man, to induce a wide range of enzymes involved in general detoxification processes of the body (McDanell *et al*., 1988[32]), and also to have a very significant effect on reducing the numbers of tumours caused by a range of carcinogens, including 3,4-benzpyrene, 1,2-dimethyl-hydrazine, and aflatoxin B_1. This has led to the suggestion that such indoles have a part to play in the protection which brassica vegetables offer against human gastric and other cancers. The second property of these indoles is their ability to react chemically with nitrite to form non-volatile nitrosamines, known to be mutagenic and considered to act as carcinogens in animals and man (Wakabayashi *et al*., 1985[33]). Studies are in progress to determine the relevance of such findings to the human condition and it is possible that future epidemiological investigations will examine in more detail than hitherto the possibility of interactions between nitrate, the precursor of nitrite, and nitrosatable indoles, such as are present in brassicas (and also some legumes).

These findings clearly indicate that, depending upon one's perspective, natural compounds in the diet may be beneficial and deleterious; what is

[29] P. A. Dahlberg, A. Bergmark, L. Björck, A. Bruce, L. Hambraeus and O. Claesson, *Am. J. Clin. Nutr*., 1984, **39**, 416.

[30] M. McMillan, E. A. Spinks, and G. R. Fenwick, *Human Toxicol*., 1986, **5**, 15.

[31] K. Sones, R. K. Heaney, and G. R. Fenwick, *J. Sci. Food Agric*., 1984, **35**, 712.

[32] R. McDanell, A. E. M. McLean, A. B. Hanley, R. K. Heaney and G. R. Fenwick *Food Chem. Toxicol*., 1988, **26**, 59.

[33] K. Wakabayashi, M. Nagao, T. Tahira, H. Saito, M. Katayama, S. Marumo, and T. Sugimura, *Proc. Jpn. Acad.*, 1985, **61B**, 190.

thus needed is to gain as wide a knowledge of the biological effects as possible so that these may be included in any risk assessment or dietary benefit analysis. In many cases it is likely to be the *overall balance* of biological activity which will be important. This is also exemplified by the saponins, considered above and briefly below.

5.3 Saponins

The intestine-permeabilizing effects of many saponins, noted above, raises the question of interactions between different classes of antinutrients or toxicants. Thus saponins are found amongst widely eaten legumes (Price *et al.*, 1988[34]), exactly those foods which society is being urged to consume because of their high fibre and protein contents. These legumes contain a wide range of additional toxicants, including tannins, oligosaccharides, phytate, lectins, enzyme inhibitors, and oestrogenic isoflavones; the effect of saponins may thus be to increase the absorption of such compounds and thereby increase their biological activity. Some evidence in support of synergistic mechanisms may be found in the work of Alvarez and Torres-Pineda (1982),[35] who observed the gut permeability of soyasaponins (Figure 4) and lectins in an *in vitro* rabbit mucosal system to be greatly enhanced in combination. A similar finding has been recently reported for the glycoalkaloids α-chaconine and α-solanine (Gee *et al.*, 1989[19]). If such effects do occur in man, and it has to be emphasized that clinical investigations have yet to be completed, then particular consideration needs to be taken of minority groups, such as vegetarians, whose intake of legumes and associated toxicants is considerably greater than that of the population at large (Ridout *et al.*, 1988[26]).

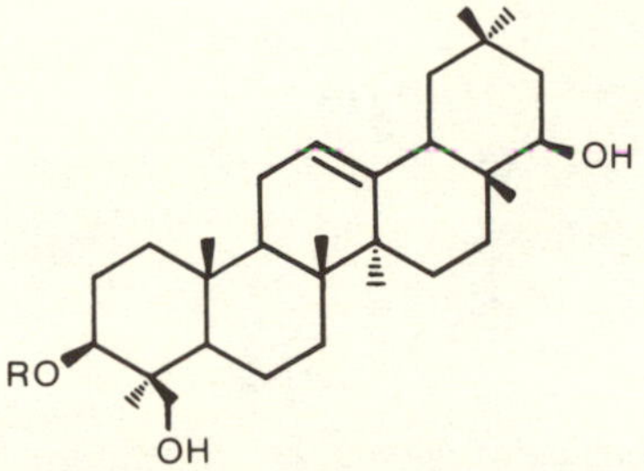

R = rha$(1\rightarrow2)$gal$(1\rightarrow2)$glu A$(1\rightarrow)-$

R = rha$(1\rightarrow2)$ara$(1\rightarrow2)$glu A$(1\rightarrow)-$

R = gal$(1\rightarrow2)$glu A$(1\rightarrow)-$

R = ara$(1\rightarrow2)$glu A$(1\rightarrow)-$

Figure 4 *Structures of typical saponins from soya*

[34] K. R. Price, J. Lewis, G. M. Wyatt, and G. R. Fenwick, *Nahrung*, 1988, **32**, 609.
[35] J. R. Alvarez and R. Torres-Pineda, *Pediatr. Res.*, 1982, **16**, 728.

The presence in soya, and its products particularly, of large amounts of isoflavones and coumestans having obvious oestrogenic effects (Price and Fenwick, 1985[12]) may also be significant in this respect. Preliminary studies (Jones *et al.*, 1989[36]) have shown that whereas the population at large has a mean daily intake of such compounds below 1 mg, that of vegetarians (who currently number some 13% of women aged between 16 and 23) may be much higher (100 mg or more). Given the findings of Axelson *et al.*, (1984)[20] and the more recent work of Setchell and co-workers, which demonstrates that such compounds can have significant effects on mammalian reproductive physiology (Setchell *et al.*, 1987[37]), preliminary clinical or epidemiological investigations of subjects consuming large amounts of soya products would appear prudent.

5.4   Illudane Sesquiterpenes

It may come as something of a surprise to learn that bracken, that most natural of countryside plants, is highly toxic (Fenwick, 1988[4]). Numerous animal studies have demonstrated chronic and acute symptoms and there is persuasive epidemiological evidence of a link between gastric and oesophageal cancers in Japan (where young bracken fronds are considered a particular delicacy) and in regions of Central and South America (where the vector is considered to be the consumption by rural populations of milk from goats and cattle foraging on the plant). There have been extensive studies over the past forty years into the biological effects of bracken and in some cases contradictory results have been obtained. It now appears that the major toxic principle, in the plant itself, is the carcinogen, ptaquiloside (Figure 5), the levels of which may be greatly affected by geographical, seasonal, environmental and genetic variables.

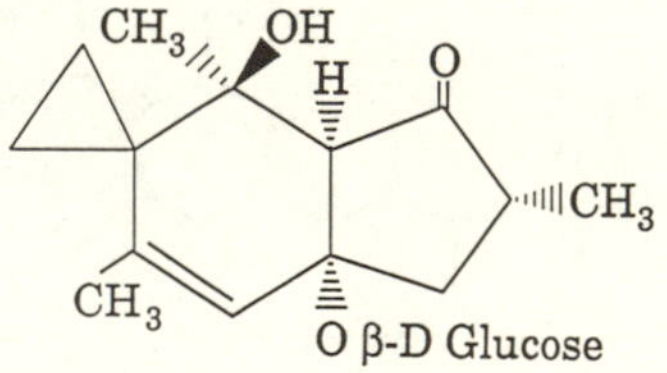

Figure 5 *Structure of the bracken carcinogen, ptaquiloside*

It is significant that in bracken research the knowledge of biological and toxicological effects (see Figure 1) is greatly in excess of the analytical information available, this being rather unusual in natural toxicant research. This lack of both a reliable analytical method and the resulting compositional data made it difficult if not impossible to compare the results

[36] A. E. Jones, K. R. Price and G. R. Fenwick, *J. Sci. Food Agric.*, 1989, **46**, 357.
[37] K. D. R. Setchell, S. J. Gosselin, M. B. Welsh, J. O. Johnston, W. F. Balistreri, L. W. Kramer, B. L. Dresser, and M. J. Tarr, *Gastroenterology*, 1987, **93**, 225.

of different laboratories. It is probable that, had a chemical assay or screening method been developed earlier, many of the problems still confronting researchers in this area would have been, at least partly, resolved.

There should be little cause for concern in the UK over the transfer of bracken carcinogens into milk and dairy products if these are obtained from cattle and processed centrally. First, the economics of conventional dairy farming are such that animals are raised on highly cultivated pasture where bracken is discouraged. Secondly, except under adverse environmental conditions (*e.g.* severe drought), cattle do not usually consume bracken in preference to other plants (although there are reports of cattle consuming bracken fronds, perhaps as a roughage, when the usual pasture is excessively lush). Once again there may be a risk if isolated communities, especially those espousing a simple, 'back to nature' philosophy, consume milk from individual cows or small numbers of cattle when feeding is not subject to control and if the milk is not bulk-processed and heat-treated. The raising of goats and the sale of goat's milk and products have increased substantially in recent years. Thus there are now more than 100 000 goats in the UK, the number in registered holdings having doubled between 1975 and 1983 (MacCarthy, 1988[38]). Such animals are usually kept on marginal land where bracken can occur. As there seem to be no reports about the effect of bracken on goats and the extent, if any, to which bracken carcinogens may be transferred to the milk, such studies would be very valuable.

Although the consumption of bracken fronds would not seem to pose a threat to human health in the UK, the recent finding by I. A. Evans and co-workers[39] that bracken spores are carcinogenic requires further study. The suggestion that farm-workers, foresters, and others regularly venturing into large areas of bracken during the period of sporulation should be encouraged to wear face masks does not seem an over-reaction.

6   Options for the Control of Natural Toxicants

The main options for the reduction of the hazard associated with the presence of natural toxicants in food and feeding stuffs are discussed in this section.

6.1   Elimination of Foods Containing Such Toxicants from the Diet

Providing that animals are fed under well defined, controlled, conditions the elimination of toxic ingredients from the ration should be relatively

[38] D. MacCarthy, 'Prodfact – 1988; A Comprehensive Guide to British Agricultural and Horticultural Produce', British Farm Produce Council, 1988, pp. 212–217.
[39] I. A. Evans, in 'Bracken, Ecology, Land Use and Control Technology', ed. R. T. Smith and J. A. Taylor, Parthenon Publishing Group, Carnforth, Lancs., 1986, pp. 139–146.

straightforward. Thus the apparently unrelated problems of egg taint and liver haemorrhage in laying hens have been solved in the UK by removing rapeseed meal from commercial layer rations (Fenwick and Curtis, 1980[40]). Where animals are free to forage then dietary control is much less easy, and good range or pasture management is essential.

The elimination of certain materials from the human diet poses considerably more of a problem, even for countries possessing centralized food production and distribution systems. State intervention can be effective, but only when supplemented by programmes of public education and when appropriate alternative foodstuffs are freely available. The lima bean contains cyanogenic glycosides, compounds which yield hydrogen cyanide upon hydrolysis. Levels of these glycosides in certain varieties are equivalent to, or greater than, 2×10^3 p.p.m. HCN, and, given the toxicity of this product, certain countries have legislated against the importation of beans containing levels greater than 200 p.p.m. (Conn, 1973[11]).

Decisions to avoid individual foods may also be made at the personal level. Thus certain people of Mediterranean extraction avoid legumes containing vicine or convicine since consumption leads to the onset of a haemolytic syndrome, favism (Chevion *et al.*, 1983[5]); people, having experienced migraine-like symptoms, avoid items such as bananas, cheese, and chocolate, which may contain the causative agents (Lovenberg, 1973[41]) and individuals prone to flatulence avoid foods which they find produce intestine gas (Price *et al.*, 1988[34]). These are all, of course, examples both of the varying composition of food products and also of the heterogeneity of the human population.

6.2   Exploitation of Normally Occurring Protective Factors in the Diet

The consumption of foods rich in dietary fibre has been suggested to offer protection against a wide range of Western diseases, including diverticular disease, colon cancer, diabetes, and heart disease (Trowell and Burkitt, 1981[42]). Much research has been conducted subsequently into understanding the mechanisms of these protective effects and into identifying sources of fibre particularly suited to alleviating individual ailments. Naturally-occurring phenolics exhibit antimutagenic and anticarcinogenic activity (Stich and Rosin, 1981[43]), onion and garlic oils are proven to be active against ischaemic heart disease and gastric cancer (Fenwick and Hanley,

[40] G. R. Fenwick and R. F. Curtis, *J. Sci. Food Agric.*, 1980, **31**, 515.
[41] W. Lovenberg, in 'Toxicants Occurring Naturally in Foods', ed. F. M. Strong, National Academy of Sciences, Washington, DC, 1973, pp. 170–188.
[42] H. C. Trowell, and D. P. Burkitt, in 'Western Diseases: Their Emergence and Prevention', ed. H. C. Trowell, and D. P. Burkitt, Edward Arnold, London, 1981, pp. 436–443.
[43] H. F. Stich, and M. P. Rosin, in 'Nutritional and Toxicological Aspects of Food Safety', ed. M. Friedman, Plenum Press, New York, 1981, pp. 1–29.

1985[44]; You *et al.*, 1988[45]), and certain glucosinolates, rich in the UK diet, are currently being examined for their anticarcinogenic and detoxifying properties (McDanell *et al.*, 1987[46]).

In general, however, although these studies are attracting increasing interest, amongst both scientists and the general public, considerable efforts will be needed before the effects in man are clinically proven, the possibility of undesirable side-effects is excluded, and dietary recommendations are proposed.

6.3 Development of Processes for Removing or Reducing Toxicants

Man has been adept at detoxifying food for many millenia. In addition to generating desirable flavour and texture, cooking is an effective detoxification process. The very first processes may well have developed from natural circumstances or accidents, such as fermentations or soaking. The relation between toxicity and bitter taste also served early man, the tongue functioning both as a bioassay and as an instrument of quality control. Many species of yam, cucurbit, cassava, and bean are inedible (due to excessive bitterness and toxicity) without prior processing. Quinoa, a protein-rich material, is potentially of considerable value to indigenous populations of South America, but the presence of toxic saponins means that prior processing is obligatory (Risi and Galwey, 1984[47]). Bitter varieties of cassava (manioc) are highly toxic because of the presence of cyanogenic glycosides and many deaths have been reported. Local techniques have been developed for processing cassava, including fermentation, steeping, and wet pounding, before consumption (Brouk, 1975[48]). It is vital that such detoxification processes take account of the local conditions and available, rural technology rather than unrealistic state-of-the-art 'first world' methodology.

An understanding of the origin, nature, and mode of action of natural toxicants may also facilitate their control. A knowledge that certain chemicals are accumulated as a response to microbial, insect, chemical, or mechanical damage will, hopefully, facilitate the establishment of agronomic, harvesting, and storage conditions designed to minimize such damage. A knowledge of the structures of individual toxicants, and of their chemical reactivities, may serve to suggest site-specific means of detoxification. Thus ammonia has been used to reduce the level of glucosinolates in

[44] G. R. Fenwick and A. B. Hanley, *CRC Crit. Rev. Food Sci. Nutr.*, 1985, **23**, 1.

[45] W. C. You, W. J. Blot, Y. S. Chang, Z. A. G. Ershow, Z. T. Yang, Q. An, B. Henderson, G. W. Xu, J. F. Fraumeni, and T. G. Wang, *Cancer Res.*, 1988, **48**, 3518.

[46] R. McDanell, A. E. M. McLean, A. B. Hanley, R. K. Heaney, and G. R. Fenwick, *Food Chem. Toxicol.*, 1987, **25**, 363.

[47] J. C. Risi and N. W. Galwey, *Adv. Appl. Biol.*, 1984, **10**, 145.

[48] B. Brouk, 'Plants Consumed by Man' Academic Press, London, 1975, pp. 96–97.

oilseed meals (Keith and Bell, 1982[49]) and ammonia and sulphur dioxide have been used to reduce the level of aflatoxins in peanut and corn products (Brekke *et al.*, 1979[50]). Care must be taken to ensure that the detoxified materials are not nutritionally inferior to the original product, especially when heat or extraction procedures are involved. Another factor, sometimes overlooked, is that the removal of one, known, toxicant may result in the formation of other, possibly unknown, biologically active species. Thus the detoxification of rapeseed or crambe meals with ferrous sulphate removes glucosinolates but results in the formation of significant amounts of nitriles, the toxicities of which have been well established (Heaney and Fenwick, 1988[27]; Mawson *et al.*, 1989[28]).

Although processing is clearly important, care must be taken to ensure that the introduction of new technology or even minor modifications to the processing conditions does not adversely affect the detoxification efficiency. Changes in culinary practice in the home may also have unforeseen consequences. Although it has long been known that beans contain toxic haemagglutins (lectins), the effects of these in human beings are minimal because of the heat treatments which precede consumption. There has recently been a trend, amongst certain sections of the population in the UK and elsewhere, to include partially cooked beans (especially red kidney beans) in salads and, since the haemagglutin has not been inactivated, a number of cases of 'food poisoning' have resulted (Bender and Reaidi, 1982[51]).

6.4   Introduction of New Plant Cultivars Containing Reduced Levels of Toxicants

Examination of the fresh produce currently available at retail outlets gives an indication of the manifold successes of the plant breeder. Improvements in yield, agronomic and harvesting characteristics, disease resistance, and tolerance to unfavourable conditions are all major objectives. In comparison much less emphasis has been placed upon improvement in food safety. This is partly due to the lack of pressure from legislators, consumer organizations, and industry. Where such pressures have been applied, *e.g.* over potato glycoalkaloids and rapeseed glucosinolates, breeders have responded. Since in many cases the presence of these biologically active species appears to be linked to plant resistance or defence, their mere removal does not necessarily produce an agronomically and commercially acceptable product and alternative protection factors may need to be introduced. The availability of a rapid, simple, and specific method for the

[49] M. O. Keith and J. M. Bell *Can. J. Anim. Sci.*, 1982, **62**, 547.
[50] O. L. Brekke, A. J. Peplinski, G. W. Nofsinger, H. F. Conway, A. C. Stringfellow, R. R. Montgomery, R. W. Silman, V. E. Johns and E. B. Bagley, *Trans. A.S.A.E.*, 1979, **22**, 425.
[51] E. A. Bender, and G. B. Reaidi, *J. Plant Foods*, 1982, **41**, 15.

screening of many thousands of plants, ideally in the early stages of growth, is a prerequisite for an effective breeding programme.

Consumer concern and attitudes towards the use of synthetic agrochemicals in intensive farming systems have resulted in a limited, but significant, move toward reduced-input farming systems. In order to grow crops successfully under wholly organic conditions, or in systems utilizing reduced agrochemical inputs, breeders and pathologists are actively seeking new plant varieties exhibiting improved 'natural' resistance. Such resistance may be associated with the presence of secondary metabolites and/or the ability effectively to synthesize and accumulate biologically active phytoalexins in response to physiological stress or damage. It is important to consider the possible consequences of such selection or plant compositional manipulation on human health and wellbeing, especially in the long term. For this reason it is vital that 'improved' varieties are not only screened for agronomic or processing characteristics, but that due consideration is also paid to biological activity (Ames, 1989[52]), difficult though this may be.

7 Conclusion

Contrary to public perceptions, artificial substances in the diet are no more hazardous than those occurring naturally and may be significantly less so. At present there have been relatively few rigorous comparisons of artificial and natural compounds, one reason for which being the absence of appropriate, simple, and inexpensive methodologies. In identifying the difficulties inherent in such work and in food toxicology generally it is important that these are not overstated and raised to the level of impossibilities. It is reasonable that equivalent attention be paid to natural toxicants, and their metabolites, as is currently given to additives, contaminants, and pesticides. The biological activity of the compound under investigation should be the major criterion, not whether it is natural or synthetic. Decisions on priorities must necessarily take account of local and national eating habits, intake, indicators of dietary trends or fashions, actual or potential toxicity, and developments in processing technology and 'novel' food materials.

In a perceptive article, Doull (1981)[53] has emphasized the importance of public education in food safety. Too much of the public's current 'education' is dependant upon alarmists, attention-seekers, and those with a vested interest in manufacturing and selling 'health foods' (which are commonly worthless and occasionally dangerous). For public acceptance and success, Doull concludes, any new food safety policy, embracing the consideration of natural toxicants, must encompass a mechanism for explaining, simply but concisely, the basis of food safety to consumers.

[52] B. N. Ames, in 'Important Advances in Oncology', J. B. Lippincott Co., Philadelphia, 1989.
[53] J. Doull in 'Food Safety', ed. H. R. Roberts, Wiley Interscience, New York, 1981, pp. 295–316.

In any event there may be pressing socioeconomic reasons which will limit the effectiveness of the programme. It is the poorest sections of society who are unable to afford the luxury of selecting their diet, of choosing high quality produce, and of discarding damaged items before cooking; in such cases problems of natural toxicants are probably greatest because the ingestion of toxicants is highest and because the toxic effects aggravate existing symptoms of malnourishment. For such people, the cutting away of damaged or infected areas of the potato represents not a minimizing of hazard but a loss of food. Until the underlying socioeconomic ills are dealt with, problems of natural toxicants exacerbated by ill health and starvation will remain.

As part of an overall programme of public health protection, natural toxicant research has an international dimension, benefitting from the exchange of ideas and data, technological and scientific collaboration, and commissioning of data bases. Public education, rather than scientific dogma, is a crucial factor in the implementation of an effective food safety programme.

Acknowledgements

The authors are grateful to Professor R. F. Curtis, formerly Director of the AFRC Institute of Food Research, to Dr H. W.-S. Chan, and members of the Bioactive Components Group, IFRN, for help in preparing this paper and to MAFF, with whose support much of the work described has been conducted.

CHAPTER 2

Polychlorinated Dibenzo-p-dioxins, Polychlorinated Dibenzofurans, and the Food Chain

JAMES R. STARTIN

1 Introduction

Polychlorinated dibenzo-p-dioxins (PCDDs) (1) and polychlorinated dibenzofurans (PCDFs) (2) are environmental contaminants which are lipophilic, chemically stable, of low volatility, and which are known to be present, even if only at very low concentrations, in the fatty tissues of animals and humans. The high toxicity which some of these compounds exhibit in animal tests has resulted in concern about the possible health consequences of this ubiquitous exposure.

(1)   (2)

For some time the term 'dioxin' was associated solely with 2,3,7,8-tetra-chlorodibenzo-p-dioxin (2,3,7,8-TCDD) but this is only one of a large number of PCDD and PCDF congeners. Since both PCDDs and PCDFs can have from one to eight chlorine substituents there is much scope for positional isomers, as shown in Table 1, and the total number of PCDD and PCDF congeners is, respectively, 75 and 135. The toxicity of different congeners varies enormously and it is unfortunate that much of the public debate which has taken place in recent years has been concerned with 'dioxins and furans', or just with 'dioxins' as a generic term for both

21

Table 1 *The total number of isomers in each PCDD and PCDF homologue and the number with lateral substitution (with substitution at each of the 2-, 3-, 7-, and 8-positions)*

Number of chlorine substituents	Number of isomers			
	PCDD		PCDF	
	Total	Laterally substituted	Total	Laterally substituted
1	2		4	
2	10		16	
3	14		28	
4	22	1	38	1
5	14	1	28	2
6	10	3	16	4
7	2	1	4	2
8	1	1	1	1
Total	75	7	135	10

groups, without regard to the particular congeners concerned.

The beginnings of concern about the possible human health effects of PCDDs can be traced to a 1970 publication about the teratogenicity of 2,4,5-trichlorophenoxyacetic acid (2,4,5-T) where a footnote added in proof attributed the observed effects to the presence of 2,3,7,8-TCDD as a contaminant.[1] Subsequently other animal tests also revealed high acute oral toxicity and carcinogenicity. Public concern and scientific activity were increased by the realization that 2,3,7,8-TCDD had been present at significant concentrations in the 'Agent Orange' defoliant used by US forces in Vietnam, and by incidents such as the use of dioxin contaminated oils for dust control in horse arenas in Missouri and in the town of Times Beach. The explosion at Seveso, Italy in 1976 at a 2,4,5-trichlorophenol production facility led to the release of an estimated 2–3 kg of 2,3,7,8-TCDD and contributed substantially to concern and scientific effort, but had been preceded by a number of other, less well publicized, industrial accidents in other locations.

During measurement of the concentrations of TCDD present in tissue from people thought to have been exposed in such incidents, it was discovered that even control samples often contained measurable concentrations of 2,3,7,8-TCDD and other PCDDs and PCDFs, and these compounds are now recognized as being ubiquitous contaminants of the environment and of human tissues. Inhalation and direct contact can only account for a very small proportion of the typical human body burden, and the major part is thought to derive from intakes associated with food.

[1] K. D. Courtney, D. E. Gaylor, M. D. Hogan, H. L. Falk, R. R. Bates, and I. Mitchel, *Science*, 1970, **168**, 864.

2 Primary Sources

Unlike many other environmental pollutants, PCDDs and PCDFs have never been manufactured deliberately, other than on the laboratory scale. A variety of primary sources are now known which contribute to environmental contamination and thus ultimately to contamination of foods. As noted above, the early concern was related to the formation of traces of 2,3,7,8-TCDD during manufacture of 2,4,5-trichlorophenol, leading to contamination in subsequent products such as 2,4,5-T and hexachlorophene. Still-bottom wastes from the manufacturing process were also often contaminated and appear to have been responsible for some of the pollution episodes in the USA, such as that at Times Beach. Controls were introduced in the UK in 1970 and 2,4,5-T has not been produced in Europe since 1983. Products containing this chemical are now of very limited availability and although environmental levels of 2,3,7,8-TCDD may still be partly attributable to past use of 2,4,5-trichlorophenol-based chemicals they are unlikely to be a significant current source.

Other chlorophenols, such as pentachlorophenol (PCP), may also be contaminated with PCDDs and PCDFs, although this principally involves the more highly chlorinated PCDD congeners; results from the analysis of some wood preservative formulations containing PCP have recently been published.[2] Polychlorinated biphenyls (PCBs) can become contaminated with PCDFs as a result of degradation during use and there have also been a number of fires involving PCB-filled equipment in which PCDFs have been formed. A broad range of chlorination levels and substitution patterns are normally involved. PCBs are no longer produced but much old electrical equipment containing PCBs continues in use and presents a risk of environmental pollution.

The presence of PCDDs and PCDFs in the fly ash of a municipal incinerator was reported in 1977,[3] and subsequent work has confirmed that emission of PCDDs and PCDFs is an almost inevitable consequence of such incineration.[4] Chemical waste incinerators, especially when used to destroy PCBs, are also potential sources as are hospital and other incinerators. Particular concern has recently been aroused by findings of high levels of PCDFs associated with wire reclamation plants.[5,6] Crummett[7] has suggested that any combustion process operating in the presence

2 W. Christmann, K. D. Klöppel, H. Partscht, and W. Rotard, *Chemosphere*, 1989, **18**, 861.
3 K. Olie, P. L. Vermeula, and O. Hutzinger, *Chemosphere*, 1977, **8**, 455.
4 F. W. Karasek and O. Hutzinger, *Anal. Chem.*, 1986, **58**, 633A.
5 A. Riss, H. Hagenmaier, U. Weberruss, C. Schlatter, and R. Wacker, 'Comparison of PCDD/PCDF Levels in Soil, Human Blood and Spruce Needles in an Area of PCDD/PCDF Contamination through Emissions from a Metal Reclamation Plant', Presented at 8th International Conference on Dioxins and Related Compounds, Umeå, 1988.
6 A. Riss and H. Hagenmaier, 'Environmental Monitoring of PCDD/PCDF in the Vicinity of a Metal Reclamation Plant in the Tyrol/Austria', Presented at 9th International Conference on Dioxins and Related Compounds, Toronto, 1989.
7 W. B. Crummett in 'Chlorinated Dioxins and Related Compounds – Impact on the Environment', ed. O. Hutzinger, R. W. Frei, E. Merian, and F. Focchiari, Pergamon Press, Oxford, 1982, p. 253.

of low concentrations of either inorganic or organic chlorides can be expected to produce traces of PCDDs and PCDFs – the 'trace chemistry of fire' hypothesis. There is increasing evidence for the *de novo* synthesis of these compounds in the absence of obvious precursors. Other combustion-related sources of PCDDs and PCDFs include exhaust emissions from vehicles fuelled by leaded petrol which normally contains chlorinated additives such as dichloroethane.[8–10] In contrast to chemical manufacturing sources, combustion processes produce almost all of the possible PCDD and PCDF congeners; the complexity of the resulting mixtures is illustrated in Figure 1. It should be noted that 2,3,7,8-TCDD typically comprises around 5% of the total TCDD produced.

A more recently discovered source which has attracted considerable attention is the chlorine bleaching of wood-pulp for paper making.[11–13] This leads to a characteristic pattern of TCDF isomers dominated by the 2,3,7,8- and 1,2,7,8-isomers with lesser amounts of many others. In addition 2,3,7,8-TCDD is normally present. This source has been of major concern in the Scandinavian countries and North America where pulp and paper making are practised on a large scale. Although there has been some attention to the contamination of foods, such as milk sold in cartons, by leaching from bleached pulp based packaging (see Section 5.4), and even over human exposure through direct contact with paper and other materials, it is the role of this source in contaminating the aquatic environment that is the main issue.

Some metallurgical processes have also been shown to act as sources, especially of PCDFs.[14] A plant in Norway where pellets of MgO mixed with coke are heated to about 700–800 °C in a chlorine atmosphere is estimated to produce an annual emission of 500 g 2,3,7,8-TCDD equivalents[15] (see Section 3.1 for a discussion of toxic equivalents calculations). In the past a nickel refining process involving high-temperature treatment of $NiCl_2$ is also thought to have produced significant amounts of TCDFs, although the current low-temperature process leads to relatively low emission rates.[15]

---

[8] K. Ballschmitter, H. Buchert, R. Niemczyk, A. Munder, and M. Swerev, *Chemosphere*, 1986, **15**, 901.

[9] S. Marklund, C. Rappe, M. Tysklind, and K. Egeback, *Chemosphere*, 1987, **16**, 29.

[10] S. Marklund, R. Andersson, M. Tysklind, C. Rappe, K.-E. Egebäck, E. Björkman, and V. Grigoriadis, *Chemosphere*, 1990, **20**, 553.

[11] US Environmental Protection Agency, 'The National Dioxin Study, Tiers 3, 5, 6 and 7', EPA 440/4-87-003, Office of Water Regulations and Standards, Washington, DC, 1987.

[12] S. E. Swanson, C. Rappe, J.Malmstrom, and K. P. Kringstad, *Chemosphere*, 1988, **17**, 681.

[13] G. Amendola, D. Barna, R. Blosser, L. LaFleur, A. McBride, F. Thomas, T. Tiernan, and R. Whittemore, *Chemosphere*, 1989, **18**, 1181.

[14] S. Marklund, L.-O. Kjeller, M. Hansson, M. Tysklind, C. Rappe, C. Ryan, H. Collazo, and R. Dougherty, in 'Chlorinated Dioxins and Dibenzofurans in Perspective', ed. C. Rappe, G. Choudhary, and L. Keith, Lewis Publishers, Michigan, 1986, p. 79.

[15] M. Oehme, S. Mano, and B. Bjerke, *Chemosphere*, 1989, **18**, 1379.

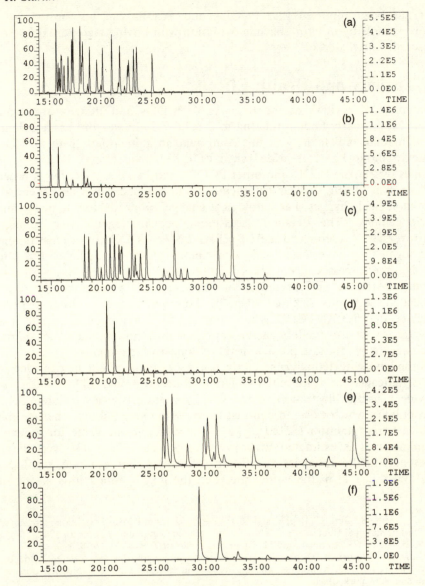

Figure 1 *Chromatograms for PCDDs and PCDFs having from 4 to 6 chlorine substituents in an extract of incinerator fly-ash. Analysis using a CPSil88 open-tubular GC column (Chrompack Ltd.) and mass spectrometric detection (selected ion monitoring): (a) TCDF (m/z 306); (b) TCDD (m/z 322); (c) PeCDF (m/z 338); (d) PeCDD (m/z 354); (e) HxCDF (m/z 374); (f) HxCDD (m/z 390)*

It has been reported that PCDDs and PCDFs can be formed from chlorophenols by *in vitro* enzymic oxidation with bovine lactoperoxidase or horseradish peroxidase.[16]

3 Toxicity and Health Effects

A number of reviews of the toxicity of PCDDs and PCDFs are available.[17-19] An early finding relating to 2,3,7,8-TCDD was the extraordinarily high acute oral toxicity of this compound in guinea-pigs, with an LD_{50} value of $0.6 \mu g kg^{-1}$. A wide range of other toxic effects is caused in animals by 2,3,7,8-TCDD and other PCDDs and PCDFs. The effects noted include skin effects such as chloracne, organ damage (liver and thymus), immunological changes, haematological changes, reproductive toxicity, and carcinogenicity. The detailed assessment of carcinogenic risk remains controversial. According to the US EPA, 2,3,7,8-TCDD is the most potent carcinogen they have tested with an estimated dose associated with a 1 in 10^6 lifetime excess cancer risk of 0.006 pg/kg bodyweight/day,[20] but a recent alternative interpretation of the same data assesses the dose associated with this risk as 0.1 pg/kg bodyweight/day.[21] Other experts conclude that 2,3,7,8-TCDD is a non-mutagenic carcinogen for which it is appropriate to use a safety factor approach in extrapolating from animal carcinogenicity data to a possible level of concern for humans.[22]

Although acute lethality is clearly not a pertinent effect in considering the significance of trace environmental concentrations, LD_{50} values do provide a useful illustration of the very large inter-species differences in susceptibility which exist. In contrast to the value given above, that for the Golden Syrian hamster is 5051 $\mu g kg^{-1}$ and although the value for humans is unknown it is estimated to be very large. There are also very large differences in the potencies of different congeners. For example, the LD_{50} for 1,3,6,8-TCDD in the guinea-pig is in the $g kg^{-1}$ range; this is one of

[16] L. G. Öberg, B. Glas, C. Rappe, and K. G. Paul, 'Biogenic Dioxin Formation', Presented at 9th International Conference on Dioxins and Related Compounds, Toronto, 1989.

[17] S. A. Skene, I. C. Dewhurst, and M. Greenberg, *Human Toxicol.*, 1989, **8**, 173.

[18] 'Halogenated Biphenyls, Terphenyls, Naphthalenes, Dibenzodioxins and Related Products', 2nd Edn., ed. R. D. Kimbrough and A. A. Jensen, Elsevier Science Publishers, Amsterdam and New York, 1989.

[19] 'Polychlorinated Dibenzo-p-dioxins and Dibenzofurans', Environmental Health Criteria No. 88, World Health Organization, Geneva, 1989.

[20] US Environmental Protection Agency, 'Health Assessment Document for Polychlorinated Dibenzo-p-dioxins and Polychlorinated Dibenzofurans', EPA/600/8-84/014F, Office of Health and Environmental Assessment, Washington, DC, 1985.

[21] R. E. Keenan, R. J. Wenning, A. H. Parsons, and D. J. Paustenbach, 'A Re-evaluation of the Tumor Histopathology of Kociba *et al.* (1978) Using 1990 Criteria: Implications for the Risk Assessment of 2,3,7,8-TCDD Using the Linearized Computer Multistage Model', Presented at 10th International Conference on Organohalogen Compounds, Bayreuth, 1990.

[22] Statement by the Committee on the Toxicity of Chemicals in Food, Consumer Products and the Environment on the Human Health Hazards of Polychlorinated Dibenzo-p-dioxins and Polychlorinated Dibenzofurans. Included in 'Dioxins in the Environment – Report of an Interdepartmental Working Group' (Pollution Paper No. 27), HMSO, London, 1989.

the dominant components in the TCDD isomer pattern from combustion sources.

Although some of the different congeners may give rise to different biological effects and act through different mechanisms, it is now accepted that consistent effects and mechanisms of action are associated with the PCDD and PCDF congeners in which each of the 2-, 3-, 7-, and 8-positions are substituted – sometimes referred to as the laterally substituted congeners. There are in all 17 compounds (see Table 1) which satisfy this structural criterion and which give rise to similar effects to 2,3,7,8-TCDD, although with considerably different potencies. An important related fact is that these laterally substituted congeners are preferentially absorbed or retained in the fat stores of animals and, regardless of the mix of congeners to which exposure occurs, are the only ones normally found in human and other animal tissues, as illustrated by Figure 2.

Epidemiological studies of exposed human populations are still underway, but the evidence so far available regarding human health effects is regarded as reassuring and humans seem to be among the least sensitive species.[22] Certainly some of the people exposed in the Seveso accident have very high residual tissue levels without apparent ill-effect. It seems most unlikely that the often used description of 'dioxin' as 'the most toxic chemical known' is appropriate when considering human health effects.

3.1 Toxic Equivalence Schemes

Because PCDDs and PCDFs normally exist in environmental and biological samples as complex mixtures of congeners having different toxicities, the concept of 2,3,7,8-TCDD equivalents has been introduced to simplify risk assessment and regulatory control. The actual concentrations of the different congeners are multiplied by a 'toxic equivalency factor' (TEF) and summed to produce a 'TCDD equivalent' (TEQ) concentration. The approach assumes that synergistic and antagonistic effects do not exist. Although a number of different schemes of TEF values have been used over the past decade the most widely accepted values are those of the International Toxicity Equivalency Factor (I-TEF) model proposed by the Committee on the Challenges of Modern Society.[23] These factors are shown in Table 2. Values are ascribed for all of the 17 laterally substituted congeners; for other congeners the I-TEF value is 0. The I-TEF model has been accepted for use in the UK.[22]

The Nordic scheme[24] is also widely used and differs only in the factor of 0.01 ascribed to 1,2,3,7,8-pentaCDF. In biological samples this congener

---

[23] 'International Toxicity Equivalency Factor Method of Risk Assessment for Complex Mixtures of Dioxins and Related Compounds, Pilot Study on International Information Exchange on Dioxins and Related Compounds', North Atlantic Treaty Organization/Committee on the Challenges of Modern Society, Report Number 176, 1988.

[24] 'Nordisk Dioxinriskbedömning, Rapport från en nordisk expertgrupp', Nord 1988:49, Nordisk Ministerråd, Copenhagen, 1988, p. 18.

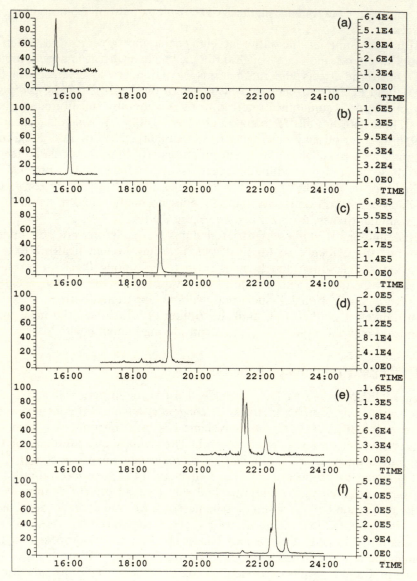

Figure 2 *Chromatograms for PCDDs and PCDFs having from 4 to 6 chlorine substituents in an extract of human milk. Analysis using a DB5 open-tubular GC column (Jones Chromatography Ltd.) and mass spectrometric detection (selected ion monitoring):* (a) *TCDF (m/z 306);* (b) *TCDD (m/z 322);* (c) *PeCDF (m/z 338);* (d) *PeCDD (m/z 354);* (e) *HxCDF (m/z 374);* (f) *HxCDD (m/z 390)*

Table 2 *The International Toxicity Equivalence Factors*

Congener	I-TEF
2,3,7,8-TCDD	1
1,2,3,7,8-PeCDD	0.5
1,2,3,4,7,8-HxCDD	0.1
1,2,3,6,7,8-HxCDD	0.1
1,2,3,6,7,8-HxCDD	0.1
1,2,3,4,6,7,8-HpCD	0.01
OCDD	0.001
2,3,7,8-TCDF	0.1
2,3,4,7,8-PeCDF	0.5
1,2,3,7,8-PeCDF	0.05
1,2,3,4,7,8-HxCDF	0.1
1,2,3,6,7,8-HxCDF	0.1
1,2,3,7,8,9-HxCDF	0.1
2,3,4,6,7,8,-HxCDF	0.1
1,2,3,4,6,7,8-HpCDF	0.01
1,2,3,4,7,8,9-HpCDF	0.01
OCDF	0.001

usually makes a very small contribution so that TEQ values calculated with the Nordic and I-TEF schemes differ very little. The scheme adopted by the German Federal Health Office assigns factors of 0.1 to 1,2,3,7,8-pentaCDD and to both 1,2,3,7,8-pentaCDF and 2,3,4,7,8-pentaCDF.[25] The difference of a factor of 5 in the weighting for the first and last of these congeners can make a substantial difference to the outcome of TEQ calculations for biological samples. Other schemes of historical and current interest are summarized in reference 23.

3.2 Guidelines for Maximum Intake

Various attempts have been made to assess acceptable levels of human exposure, or at least to assess the risk to health. In the UK the Committee on the Toxicity of Chemicals in Food, Consumer Products and the Environment has recommended[22] that a figure of 1 pg/kg bodyweight/day be considered as a guideline value in the sense that this is a level which, when exceeded, should trigger investigation and appropriate measures to reduce environmental levels generally. In expressing this opinion the committee stated that, because of the considerable uncertainties inherent in deriving quantitative assessments, this should not be regarded as a tolerable daily intake.

[25] Umweltbundesamt (UBA), Sachstand Dioxine – Stand November 1984, Erich Schmidt Verlag, Berlin.

Limits which have been proposed or introduced in other countries include: Nordic Countries, 35 pg TCDD equivalents/kg bodyweight/week;[24] Netherlands, 4 pg/kg bodyweight/day; Federal Republic of Germany 1–10 pg/kg bodyweight/day;[25] Ontario, Canada 10 pg/kg bodyweight/day.[26] As pointed out by Barnes,[27] these doses were established by considering essentially the same evidence but applying different safety factors to the experimental 'no observable adverse effect levels'.

More recently a group of experts convened by the World Health Organization/Regional Office for Europe has recommended that the tolerable daily intake be set at 10 pg/kg bodyweight/day.[28]

There are very few examples of specific regulations or advice pertaining to particular dietary commodities. In Canada a 'Virtually Safe Dose' for 2,3,7,8-TCDD in fish of 20 ng kg^{-1} was established as a tolerance value,[29] and in the Netherlands an action level for milk of 6 ng kg^{-1} TEQ has been declared.[30]

4   Environmental Pathways

The entry point for PCDDs and PCDFs to the environment, as well as the pattern of congeners concerned, varies with the different sources. In the case of combustion the most direct route is through the atmosphere, where aerial transport of vapour-phase and particulate-bound material (such as fly ash) may result in dispersion of PCDDs and PCDFs over a considerable area before they settle or are washed by rain to ground level. PCDDs and PCDFs are of low volatility so that gas-phase transport is likely to be less important than adsorption on particulates. In the case of incinerators, PCDDs and PCDFs will also be present in the solid debris, such as grate ash, and can enter the environment when this is disposed of by landfill or other means.

Most of the PCDDs and PCDFs formed during pulp bleaching appear in the liquid effluent from the process and thus enter the aqueous environment directly. Again most of the PCDDs and PCDFs are likely to be particulate bound rather than in solution.

---

[26] Ontario Ministry of the Environment, Polychlorinated Dibenzo-*p*-dioxins and Polychlorinated Dibenzofurans, Scientific Criteria Document for Standard Development, No. 4-84, Toronto, Canada, 1985.

[27] D. G. Barnes, *Chemosphere*, 1989, **18**, 33.

[28] Consultation on tolerable daily intake from food of PCDDs and PCDFs, Report of meeting of WHO expert group, Bilthoven (4–7 December 1990), EUR/ICP/PCS 030(S), 21 December 1990.

[29] H. Tosine, 'Dioxins: a Canadian Perspective', in 'Chlorinated Dioxins and Dibenzofurans in the Total Environment', ed. G. Choudhary, L. H. Keith, and C. Rappe, Butterworth, Boston, 1983, p. 3.

[30] A. K. D. Liem, R. Hoogerbrugge, P. R. Kootstra, A. P. J. M. de Jong, J. A. Marsman, A. C. den Boer, R. S. den Hartog, G. S. Groenemeijer, and H. A. van't Klooster, 'Levels and Patterns of Dioxins in Cow's Milk in the Vicinity of Municipal Waste Incinerators and Metal Reclamation Plants in the Netherlands', Presented at 10th International Conference on Organohalogen Compounds, Bayreuth, 1990.

A further distribution mechanism that may be of considerable significance results from the presence of PCDDs and PCDFs in sewage sludge.[31-33] The primary sources have not been elucidated, but presumably could include contributions from industrial effluents, from that portion of the human dietary intake which is not absorbed, and from lavatory paper made from chlorine bleached pulp. It has not been established if PCDDs and PCDFs are actually formed in sewage, but the widespread use of chlorine-based bleaches must satisfy one of the prerequisite conditions. The evidence that biogenic formation is possible may also be of relevance.[16] Sewage sludge is disposed of by landfill and controlled sea dumping but is also used on agricultural land as a soil conditioner and fertilizer. One study has demonstrated elevated levels of PCDDs and PCDFs in soil as a result of regular sewage sludge application.[34] In Germany the application of sewage sludge to agricultural land has been restricted as a result of such findings.[34]

Clearly PCDDs and PCDFs will be present in surface waters as a result of direct contamination and atmospheric deposition; in sediments; in soils as a result of atmospheric deposition and the application of chemicals and sewage sludge; on surfaces as a result of atmospheric deposition; and, of course, in the air.

The concentrations found in urban air in a number of industrialized countries have been summarized by Travis,[35] and cover a remarkably narrow range from 1.3–9.6 pg m^{-3}. In the UK information is available on the levels of PCDDs and PCDFs in soil samples collected from points on a 50 km grid, thus defining the background levels.[36,37] This makes it clear that PCDDs and PCDFs are ubiquitous in the environment and occur in significantly higher concentrations in urban areas as compared with rural locations. Background levels for the TCDD homologue had a range of <0.5–69 ng kg^{-1} with a median of 6.0 ng kg^{-1} and levels of 2,3,7,8-TCDD were found to be in the range <0.5–2.1 ng kg^{-1}. In the case of urban soils the sum of the means for each homologue group was 2155 ng kg^{-1} similar to the mean concentration of 1405 ng kg^{-1} found in urban soils in the USA.[38]

[31] L. L. Lamparski, T. J. Nestrick, and V. A. Stenger, *Chemosphere*, 1984, **13**, 361.

[32] H. Hagenmaier, H. Brunner, W. Knapp and U. Weberruss, 'Untersuchungen der Gehalte an Polychlorierten Dibenzodioxinen, Polychlorierten Dibenzofuranen und Ausgewählten Chlorkohlenwasserstoffen in Klärschlämmen', Forschungsbereich 103 03 305, Im Auftrag des Umweltbundesamtes, 1988.

[33] C. Rappe, L.-O. Kjeller, and R. Anderson, *Chemosphere*, 1989, **19**, 13.

[34] M. S. McLachlan and M. Reissinger, 'The Influence of Sewage Sludge Fertilization on the PCDD/F Concentration in Soil', Presented at 10th International Conference on Organohalogen Compounds, Bayreuth, 1990.

[35] C. C. Travis and H. A. Hattemer-Frey, *Risk Analysis*, 1989, **9**, 91.

[36] C. S. Creaser, A. R. Fernandes, A. Al-Haddad, S. J. Harrad, R. B. Homer, P. W. Skett, and E. A. Cox, *Chemosphere*, 1989, **18**, 767.

[37] H. M. Inspectorate of Pollution, 'Determination of Polychlorinated Biphenyls, Polychlorinated Dibenzo-*p*-dioxins and Polychlorinated Dibenzofurans in UK Soils', Technical Report, HMSO, London, 1989.

[38] T. J. Nestrick *et al.*, *Chemosphere*, 1986, **15**, 1453.

A number of authors have used mathematical models to predict how PCDDs and PCDFs are partitioned in the environment and to suggest the major pathways of human exposure.[39–42] These studies tend to emphasize 2,3,7,8-TCDD, for which the most comprehensive physicochemical data are available. The calculations indicate that the predominant route of human exposure is through food – Travis,[39] for example, estimates 98% – and support the intuitive judgement that fatty foods are likely to be of greatest significance. In addition to these calculations the results of laboratory analysis of foods are becoming available.

5   Accumulation and Levels in the Food Chain

5.1   Vegetation

Plants are exposed to PCDDs and PCDFs via soil, groundwater, and by direct aerial deposition. Studies on uptake from soil, which have recently been reviewed,[43] are not entirely consistent. There is some evidence that the root systems of plants can absorb TCDD from soil but that this is concentrated in the outer surfaces of the roots;[44] passive uptake, in which soil particles are incorporated into the epidermis of the root tissue, has been suggested as a probable mechanism.[45] Most of the evidence suggests that there is no translocation within plants. PCDD and PCDF levels on foliage and fruits thus represent contributions from direct deposition of air-borne particulates and by absorption of vapour phase contaminants from the air including those which are attributable to evaporation from the soil.[46]

Not surprisingly, contamination of vegetation is associated with areas where there are specific PCDD and PCDF sources. High TCDD levels ($100 \, \text{ng kg}^{-1}$) were detected in the peel of fruits grown on soils contaminated in the Seveso incident, but not in the flesh.[47] Recent results from the vicinity of a wire reclamation incinerator show considerable contamination ($5–10 \, \text{ng kg}^{-1}$ TEQ) of leaf vegetables and rather less of fruits.[48] Apart

[39] C. C. Travis and H. A. Hattemer-Frey, *Chemosphere*, 1987, **16**, 2331.

[40] K. C. Jones and B. G. Bennett, *Sci. Total Environ.*, 1989, **78**, 99.

[41] T. E. McKone and P. B. Ryan, *Environ. Sci. Technol.*, 1989, **23**, 1154.

[42] J. B. Stevens and E. N. Gerbec, *Risk Analysis*, 1988, **8**, 329.

[43] G. A. Kew, J. L. Schaum, P. White, and T. T. Evans, *Chemosphere*, 1989, **18**, 1313.

[44] S. Facchetti, A. Balasso, C. Fichtner, G. Frare, A. Leoni, C. Mauri, and M. Vasconi, in 'Chlorinated Dioxins and Dibenzofurans in Perspective', ed. C. Rappe, G. Choudhary, and L. H. Keith, Lewis Publishers Inc., Michigan, 1986, p. 225.

[45] A. L. Young, 'Long Term Studies on the Persistence and Movement of TCDD in a Natural Ecosystem', in 'Human and Environmental Risks in Chlorinated Dioxins and Related Compounds', ed. R. E. Tucker, A. L. Young, and A. P. Gray, Plenum Press, New York, 1983, p. 173.

[46] A. Reischl, M. Reissinger, H. Thoma, and O. Hutzinger, *Chemosphere*, 1989, **19**, 467.

[47] H.-K. Wipf, E. Homberger, N. Neuner, U. B. Ranalder, W. Vetter, and J. P. Vuilleumier, in 'Chlorinated Dioxins and Related Compounds – Impact on the Environment', ed. O. Hutzinger, R. W. Frei, E. Merian, and F. Pocchiari, Pergamon Press, Oxford, 1982, p. 115.

[48] B. Prinz, G. H. M. Krause, and L. Radermacher, 'Criteria for the Evaluation of Dioxins in Vegetable Plants and Soils', Presented at 10th International Conference on Organohalogen Compounds, Bayreuth, 1990.

from this localized contamination, levels of PCDDs and PCDFs in fruit and vegetables do not seem to make a significant contribution to intake although the amount of data relating to samples obtained at the retail level is limited. Although Davies[49] has published results for samples from Ontario, Canada, showing high levels in a fruit composite sample which consisted mainly of apples, more recent work with much more rigorous analytical quality control has shown levels in a variety of fruit and vegetable samples from this region to be below the limit of detection (<1 ng kg^{-1}).[50] In a survey of foodstuffs available in West Berlin five vegetable samples were analysed (including cauliflower, lettuce, cherries, and apples) and again PCDDs and PCDFs were not found, subject to a detection limit of 0.01 ng kg^{-1} for each isomer.[51] These were presumably background samples but it is not clear whether they were washed or whether outer leaves were discarded prior to analysis.

The levels of PCDDs and PCDFs in grass are also of interest since grazing animals, such as cattle and sheep, provide a link to the human diet through meat and dairy products. Preliminary results have been published[52] from a survey of levels in herbage in the UK using samples obtained from the same locations as the soil samples discussed above. The results so far available are for homologue group totals rather than specific congeners so that the toxic equivalents approach is not appropriate. In 67 samples the results for the TCDD group, for example, span the range 0.3–74.0 ng kg^{-1} dry weight with an interquartile range of 2.3–9.7 ng kg^{-1} and a median of 4.5 ng kg^{-1}; the contribution of 2,3,7,8-TCDD is estimated to be below 0.1 ng kg^{-1}. These results are clearly very similar to those cited for soils.

5.2 Animals

Grazing animals are potentially exposed to PCDDs and PCDFs through the ingestion of grass (or hay or silage), other feeds and water, and by inhalation. Intake calculations based on environmental modelling vary somewhat. Travis[39] suggests a typical daily intake by cattle of about 0.75 ng day^{-1} 2,3,7,8-TCDD, while Stevens[42] has calculated the intake of all PCDDs to be in the region of 0.15 ng day^{-1} for cattle in the vicinity of a municipal solid waste incinerator. Despite these differences in the quantitative assessment of dose both approaches lead to the conclusion that forage is the main source to the cattle. Travis estimates the contribution of forage and soil ingestion to be 80% and 20% respectively.

Laboratory studies have shown that 2,3,7,8-TCDD can be absorbed quite readily by the digestive system of animals; for rats a figure of 50–60% has been demonstrated,[53] and a study on a human volunteer

[49] K. Davies, *Chemosphere*, 1988, **17**, 263.
[50] B. Birmingham, B. Thorpe, R. Frank, R. Clement, H. Tosine, G. Fleming, J. Ashman, J. Wheeler, B. D. Ripley, and J. J. Ryan, *Chemosphere*, 1989, **19**, 507.
[51] H. Beck, K. Eckart, W. Mathar, and R. Wittkowski, *Chemosphere*, 1989, **18**, 417.
[52] J. R. Startin, M. Rose, and C. Offen, *Chemosphere*, 1989, **19**, 531.
[53] G. F. Fries and G. S. Marrow, *J. Agric. Food Chem.*, 1975, **23**, 265.

showed 87% absorption.[54] Distribution within body tissues favours fatty tissues, but lipid adjusted concentrations are usually fairly uniform. Although the dietary intake will, in most cases, include a complex mixture of congeners only the laterally substituted congeners persist in animal tissue.

5.2.1   Meat

Results from the analysis of retail samples of meat have been presented by Beck[51] and by Furst,[55] although these studies are based on very limited sample numbers. After adjusting for the differences between the I-TEF and German toxic equivalents schemes both of these studies indicate levels in beef, mutton, and chicken fat in the region of 1–5 ng kg^{-1} TEQ. It should be noted that these results are given on a fat weight basis and, since the fat content of the samples analysed is not reported, it is not possible to correct to a whole sample basis. Rather lower levels are reported for pork.

In an earlier study Ryan found a considerable incidence of hexa-, hepta-, and octa-chlorinated congeners in pork and chicken fat from various locations in Canada, associated with the use of PCP treated wood shavings as litter.[56]

5.2.2   Milk

In studies of lactating dairy cows fed a single dose of tritiated 2,3,7,8-TCDD, about 60% was eliminated in the faeces and about 13% appeared in the milk within 14 days, regardless of whether the TCDD was ingested with soil or with grain.[57,58] In an alternative approach with normal feed and under approximately steady-state conditions, McLachlan has shown that about 20% of ingested 2,3,7,8-TCDD appears in milk.[59] The proportion transferred decreased considerably with increasing chlorination level.

A number of reports have dealt with levels of PCDDs and PCDFs in cow's milk from various European countries.[30,51,55,60,61] The levels given in Table 3 for a number of farms in the UK are quite typical. In contrast a considerably lower background (0.006 ng kg^{-1} TEQ) has been reported in

[54] H. Poiger and C. Schlatter, *Chemosphere*, 1986, **15**, 1489.

[55] P. Furst, C. Furst, and W. Groedel, *Chemosphere*, 1990, **20**, 787.

[56] J. J. Ryan, R. Lizotte, T. Sakuma, and B. Mori, *J. Agric. Food Chem.*, 1985, **33**, 1023.

[57] D. Jones, S. Safe, E. Morcom, M. Holcomb, C. Coppock, and W. Ivie, *Chemosphere*, 1987, **16**, 1743.

[58] D. Jones, S. Safe, E. Morcom, M. Holcomb, C. Coppock, and W. Ivie, *Chemosphere*, 1989, **18**, 1257.

[59] M. S. McLachlan, H. Thoma, M. Reissinger, and O. Hutzinger, *Chemosphere*, 1990, **20**, 1013.

[60] C. Rappe, M. Nygren, G. Lindstrom, H. R. Buser, O. Blaser, and C. Wuthrich, *Environ. Sci. Technol.*, 1987, **21**, 964.

[61] J. R. Startin, M. Rose, C. Wright, I. Parker, and J. Gilbert, *Chemosphere*, 1990, **20**, 793.

Table 3 *Levels of PCDDs and PCDFs found in milk from rural locations in the UK (from ref. 61)*

Congener	ng kg⁻¹ whole milk		
	Min	*Max*	*Mean*
2,3,7,8-TCDD	<0.004	0.013	0.009
1,2,3,7,8-PeCDD	0.012	0.023	0.016
1,2,3,4,7,8-HxCDD/	0.019	0.043	0.032
1,2,3,6,7,8-HxCDD			
1,2,3,7,8,9-HxCDD	<0.007	0.018	0.010
1,2,3,4,6,7,8-HpCDD	<0.022	0.066	0.046
OCDD	0.215	0.323	0.230
2,3,7,8-TCDF	<0.006	0.011	0.008
1,2,3,7,8-PeCDF	<0.002	<0.017	0.005
2,3,4,7,8-PeCDF	0.028	0.038	0.032
1,2,3,4,7,8-HxCDF	0.013	0.026	0.017
1,2,3,6,7,8-HxCDF	0.009	0.017	0.012
1,2,3,7,8,9-HxCDF	<0.002	<0.012	0.008
2,3,4,6,7,8-HxCDF	0.007	0.017	0.012
1,2,3,4,6,7,8-HpCDF	<0.007	<0.067	0.020
1,2,3,4,7,8,9-HpCDF	<0.010	<0.067	0.026
OCDF	0.023	0.071	0.041
Total TEQ	0.034	0.052	0.045

milk from New Zealand where the population is sparse and where municipal waste incineration is not practised.[62] A number of studies have shown a more localized influence of incinerators and other sources. In the Netherlands TEQ levels of 5–12 ng kg⁻¹ fat have been found in the vicinity of one incinerator compared to 0.7–2.5 ng kg⁻¹ from reference locations.[30] Samples from two farms in each of two locations in the UK where there are believed to be potential sources gave TEQ levels in the range 0.12–0.25 ng kg⁻¹ whole milk.[61] The corresponding lipid adjusted range of 3.0–6.25 ng kg⁻¹ is above background although the highest level is well below the highest found in the Netherlands.

The laterally substituted congeners are also predominant in milk, but some traces of other PCDDs and PCDFs are often evident, presumably because cows ingest the full range of combustion related congeners and because of the rapid distribution of intake to milk.

It should be noted that, as shown in Table 4, 2,3,4,7,8-PeCDF makes a major contribution to the total concentration measured as toxic equivalents.

[62] S. J. Buckland, D. J. Hannah, J. A. Taucher, and R. J. Weston, 'The Migration of Polychlorinated Dibenzo-*p*-Dioxins and Polychlorinated Dibenzofurans into Milks and Cream from Bleached Paperboard Packaging', Presented at 10th International Conference on Organohalogen Compounds, Bayreuth, 1990.

Table 4 *Contribution of individual PCDD and PCDF congeners to the total toxic equivalents (TEQ) concentration in milk, based on the mean levels recorded in Table 3 and the I-TEF factors given in Table 2*

Congener	Contribution to TEQ total (%)
2,3,7,8-TCDD	20
1,2,3,7,8-PeCDD	18
1,2,3,4,7,8-/1,2,3,6,7,8-HxCDD	7
1,2,3,7,8,9-HxCDD	2
1,2,3,4,6,7,8-HpCDD	1
OCDD	1
2,3,7,8-TCDF	2
1,2,3,7,8-PeCDF	1
2,3,4,7,8-PeCDF	36
1,2,3,4,7,8-HxCDF	4
1,2,3,6,7,8-HxCDF	3
1,2,3,7,8,9-HxCDF	2
2,3,4,6,7,8-HxCDF	3
1,2,3,4,6,7,8-HpCDF	< 1
1,2,3,4,7,8,9-HpCDF	1
OCDF	< 1

5.2.3 Eggs

There is very limited information on the incidence of PCDDs and PCDFs in hen's eggs. Beck has reported results for a single sample which gave a total concentration, in I-TEQ units, of 1.61 ng kg^{-1} on a fat weight basis.[51] Results have been reported from studies in the USA which show considerably higher levels in eggs from free-range birds as compared with those from more intensive conditions, and a correlation between levels in the eggs and those in soil.[63]

5.3 Fish

The accumulation of lipophilic chemicals in fish and transfer on to fish predators, including man, is well known. Although it is almost impossible to detect PCDDs and PCDFs in water these compounds can be found in fish, even from the oceans. A number of the studies of PCDD and PCDF levels in fish have dealt primarily with areas, such as the Great Lakes, with

[63] R. D. Stephens, M. Harnly, D. G. Hayward, R. R. Chang, J. Flattery, M. X. Petreas, and L. Goldman, *Chemosphere*, 1990, **20**, 1091.

a particular pollution history.[64-66] An extensive survey of fish from inland waters in the USA was conducted as part of a general programme of environmental screening.[67]

Some results are also available indicating the levels in fish as consumed in Western Europe.[51,55,61,68,69] In the UK the results from eight retail samples, including plaice, mackerel, herring, cod, skate, and coley, gave a mean of $0.74 \, \text{g kg}^{-1}$ and a range of $0.15-1.84 \, \text{ng kg}^{-1}$ TEQ on a wet weight basis.[61] Most of the other results available are broadly similar to those from the UK with the exception of the data from Sweden,[69] which show considerably more elevated levels in fish from lakes and from the Baltic Sea and the Gulf of Bothnia, presumably as a result of the polluted status of these waters.

The selectivity of uptake found in animals also applies to fish but is rather less complete. It is thus usual for the laterally substituted congeners to be the major PCDD and PCDF contaminants present, but for smaller amounts of other congeners also to be found. With crabs and some other crustaceans there seems to be little selectivity and the relative contributions of the different congeners in tissue represent those in the exposure. This is of some significance in attempting to attribute contamination to specific sources on the basis of source-characteristic congener patterns.[70]

5.4 Leaching from Paper and Paperboard

The formation of PCDDs and PCDFs during the chlorine bleaching of wood pulp can not only cause environmental contamination, but can also lead to these compounds being present in the finished paper and board.[71-73] Studies involving coffee filters showed them to contain

[64] J. J. Ryan, Pul-Yan Lau, J. C. Piulon, D. Lewis, H. A. McLeod, and A. Gervals, *Environ. Sci. Technol.*, 1984, **18**, 719.

[65] N. V. Fehringer, S. M. Walthers, R. J. Kozara, and L. F. Schneider, *J. Agric. Food Chem.*, 1985, **33**, 626.

[66] D. De Vault, W. Dunn, P.-A. Bergqvist, K. Wiberg, and C. Rappe, *Environ. Toxicol. Chem.*, 1989, **8**, 1013.

[67] D. W. Kuehl, B. C. Butterworth, A. McBride, S. Kroner, and D. Bahnick, *Chemosphere*, 1989, **18**, 1997.

[68] A. Biseth, M. Oehme, and K. Faerden, 'Levels of Polychlorinated Dibenzo-*p*-dioxins and Polychlorinated Dibenzofurans in Selected Norwegian Foods', Presented at 10th International Conference on Organohalogen Compounds, Bayreuth, 1990.

[69] C. de Wit, B. Jansson, M. Strandell, M. Ohlsson, S. Bergek, M. Boström, P.-A. Bergqvist, C. Rappe, and Ö. Andersson, 'Polychlorinated Dibenzo-*p*-dioxin and Dibenzofuran Levels and Patterns in Fish Samples Analyzed within the Swedish Dioxin Survey', Presented at 10th International Conference on Organohalogen Compounds, Bayreuth, 1990.

[70] M. Oehme, A. Bartnova, and J. Knutzen, *Environ. Sci. Technol.*, 1990, **24**, 1836.

[71] H. Beck, K. Eckart, W. Mathar, and R. Wittkowski, *Chemosphere*, 1988, **17**, 51.

[72] V. H. Kitunen and M. S. Salkinoja-Salonen, *Chemosphere*, 1989, **19**, 721.

[73] K. Wiberg, K. Lundström, B. Glas, and C. Rappe, *Chemosphere*, 1989, **19**, 735.

[72] V. H. Kitunen and M. S. Salkinoja-Salonen, *Chemosphere*, 1989, **19**, 721.

[73] K. Wiberg, K. Lundström, B. Glas, and C. Rappe, *Chemosphere*, 1989, **19**, 735.

$1-5$ ng kg^{-1} 2,3,7,8-TCDD and $10-30$ ng kg^{-1} 2,3,7,8-TCDF.[74,75] Although the evidence suggests that a considerable proportion, 30% or more, can be extracted during coffee brewing the resulting concentrations in coffee remain very low. A risk assessment of the potential exposure to PCDDs and PCDFs associated with the consumption of tea brewed with bleached-pulp tea bags has concluded that this is negligible.[76]

It has also been shown that some leaching can occur from paperboard milk cartons into milk, despite the presence of an interior polyethylene surface coating.[77–80,61] The evidence suggests that some 10–20% of the 2,3,7,8-TCDF in the carton material is transferred to milk on storage for up to 12 days. It is rather less clear whether or not equivalent migration of 2,3,7,8-TCDD also occurs. The results from Canada[77] and New Zealand[78] suggest that this migration contributes substantially to the total concentration in the milk. In Europe, however, where the background levels in milk are somewhat greater and levels in the cartons are typically lower, migration accounts for a smaller proportion of the typical final concentrations. The introduction of improved pulp production technology, which produces markedly lower PCDD and PCDF concentrations,[81] is likely to greatly reduce this route of human exposure.

6 Levels in Human Milk and Tissue

In general, results on human tissue are remarkable more for the uniformity of the background levels than for the differences. For 2,3,7,8-TCDD, for which the greatest quantity of data is available, adipose tissue levels for individuals from industrialized countries fall in the range $3-20$ ng kg^{-1} with a mean of 7 ng kg^{-1}.[40]

One study, in Sweden, has demonstrated a clear difference between a group of fish-allergic individuals, a control group, and a group with a high

[74] H. Beck, A. Dross, K. Eckart, W. Mathar, and R. Wittkowski, *Chemosphere*, 1989, **19**, 655.

[75] National Council of the Paper Industry for Air and Stream Improvement Inc., 'Assessment of the Risks Associated with the Potential Exposure to Dioxins through the Consumption of Coffee Brewed Using Bleached Paper Coffee Filters', NCASI Technical Bulletin 546, New York, 1988.

[76] M. J. Sullivan and J. T. Stanford, *Chemosphere*, 1990, **20**, 1755.

[77] J. J. Ryan, L. G. Panopio, and D. A. Lewis, 'Bleaching of Pulp and Paper as a Source of PCDDs and PCDFs in Food', Presented at 8th International Symposium on Dioxins and Related Compounds, Umeå, 1988.

[78] S. J. Buckland, D. J. Hannah, J. A. Taucher, and R. J Weston, 'The Migration of Polychlorinated Dibenzo-*p*-dioxins and Dibenzofurans into Milks and Cream from Bleached Paperboard Packaging', Presented at 10th International Conference on Organohalogen Compounds, Bayreuth, 1990.

[79] L. LaFleur, T. Bousquet, K. Ramage, B. Brunck, T. Davis, W. Luksemburg, and B. Peterson, *Chemosphere*, 1990, **20**, 1657.

[80] C. Rappe, G. Lindström, B. Glas, and K. Lundström, *Chemosphere*, 1990, **20**, 1649.

[81] V. H. Kitunen and M. S. Salkinoja-Salonen, *Chemosphere*, 1990, **20**, 1663.

fish consumption.[82] There is, however, no suggestion of any health effect being attributable to this difference in exposure.

Human milk has been of considerable interest, partly because it is the sole food source for many infants, but also because it is a non-invasive way of looking at levels in human tissue. A unique initiative by the World Health Organization/Regional Office for Europe has made available results from many regions of the world in a study featuring a comprehensive scheme of interlaboratory calibration.[83] The average levels in milk from women living in widely dispersed locations in Western Europe were remarkably similar and in the range 30–40 ng kg^{-1} (TEQ total on a fat basis). Rather lower levels were reported in samples from some Nordic countries.

7 Analytical Considerations

As might be expected in view of the very low concentrations concerned and the complexity of the mixtures of congeners, the analysis of PCDDs and PCDFs in foods and other biological samples is challenging. Successful analysis of the minute concentrations present in biological samples and in foods inevitably requires access to a high-performance mass spectrometer and needs extensive and time-consuming sample clean-up. Many different variations of methodology are available and these have been summarized in a recent comprehensive review.[84] The use of stable isotope internal standards is now regarded as an essential step for correct measurement. Guidelines for the validation and quality control of analytical work have recently been proposed.[85]

There are a number of points which need to be clearly appreciated in assessing the reported data. The analytical standards, on which the accuracy of quantification depends, are normally available only as dilute solutions and there must be some concern as to the initial and long-term accuracy of these solutions. There are some unfortunate differences of practice in reporting results on a whole sample or lipid basis which can hamper comparison. The effects of the different toxic equivalent schemes must be considered. The level of some individual congeners are often close to the analytical detection limit and may be subject to considerable imprecision in measurement. In calculating total levels in toxic equivalent units there are differences of approach in accounting for congeners which have not been detected. In some reports the actual level is assumed to be

[82] B.-G. Svensson, A. Nilsson, M. Hansson, C. Rappe, B. Akesson, and S. Skerfving, 'Fish Consumption and Human Exposure to Dioxins and Dibenzofurans', Presented at 10th International Conference on Organohalogen Compounds, Bayreuth, 1990.

[83] 'Levels of PCBs, PCDDs and PCDFs in Breast Milk', *Environ. Health*, **34**, World Health Organization/Regional Office for Europe, Copenhagen, 1989.

[84] R. E. Clement and H. M. Tosine, *Mass Spectrom. Rev.*, 1988, **7**, 593.

[85] P. F. Ambidge, E. A. Cox, C. S. Creaser, M. Greenberg, M. G. de Gem, J. Gilbert, P. W. Jones, M. G. Kibblewhite, J. Levey, S. G. Lisseter, T. J. Meredith, L. Smith, P. Smith, J. R. Startin, I. Stenhouse, and M. Whitworth, *Chemosphere*, 1990, **21**, 999.

zero, and in others to be 50% or 100% of the detection limit. It must also be recognized that real detection limits vary between analyses and between individual congeners in a single analysis. Unless great care is taken in the interpretation of results it is possible for differences in analysis and reporting to cause apparent differences in the PCDD and PCDF levels found.

8 Conclusion

It is clear that, although food is the primary source of PCDDs and PCDFs to man, the extent to which foods contain these compounds is not a product of agricultural or food manufacturing practice but depends almost entirely on the general level of environmental contamination. Compared with the concentrations of many other man-made and natural toxins PCDDs and PCDFs are present at extremely low levels. There is no clear evidence that these small levels of exposure pose a significant risk to human health, although it is impossible to rule this out entirely. The presence of an entirely man-made contaminant in the environment and in food is undesirable but the only feasible approach to control is by the abatement or elimination of the primary sources. Nevertheless a programme of monitoring to better identify and quantify the sources and to gain further information on average and extreme intakes is likely to remain necessary.

CHAPTER 3

Analysis and Occurrence of Polycyclic Aromatic Hydrocarbons in Food

KEITH D. BARTLE

1 Nature and Biological Action of Polycyclic Aromatic Hydrocarbons

Polycyclic aromatic hydrocarbons (PAHs) comprise the largest class of known environmental carcinogens;[1] some, while not carcinogenic, may act as synergists. PAHs are found in water, air, soil, and, therefore, food; they originate from diverse sources such as tobacco smoke, engine exhausts, petroleum distillates, and coal-derived products, with combustion sources predominating.[2-4] PAHs in food may also arise from smoke curing, charcoal broiling, food additives, and packaging as well as environmental pollution.[5]

PAHs consist[6] (Figure 1) of benzene units linked together, either cata-annellated [*e.g.* anthracene (1), linearly annellated, and phenanthrene (2), angularly annellated] or peri-condensed [*e.g.* pyrene (3)]. These possibilities, taken with the number of alkylated derivatives, make the

[1] 'Environmental Carcinogens: Polycyclic Aromatic Hydrocabons', ed. G. Grimmer, CRC Press, Boca Raton, Florida, 1983.
[2] M. L. Lee. M. V. Novotny and K. D. Bartle, 'Analytical Chemistry of Polycyclic Aromatic Compounds', Academic Press, New York, 1981, Chapter 2, p. 17.
[3] A. Bjørseth and T. Ramdahl, in 'Handbook of Polycyclic Aromatic Hydrocarbons', ed. A. Bjørseth and T. Ramdahl, Marcel Dekker, New York, 1985, Vol. 2, Chapter 1, p. 1.
[4] T. Vo-Dinh, in 'Chemical Analysis of Polycyclic Aromatic Compounds', ed. T. Vo-Dinh, John Wiley, New York, 1989, Chapter 1, p. 1.
[5] T. Fazio and J. W. Howard, in 'Handbook of Polycyclic Aromatic Hydrocarbons', ed. A. Bjørseth, Marcel Dekker, New York, 1983, Vol. 1, Chapter 11, p. 461.
[6] M. Zander, in 'Handbook of Polycyclic Aromatic Hydrocarbons', ed. A. Bjørseth, Marcel Dekker, New York, 1983, Vol. 1, Chapter 1, p. 1.

number of isomers for even a modest molecular weight enormous (Table 1) and PAH mixtures from combustion *etc.* are usually extremely complex.

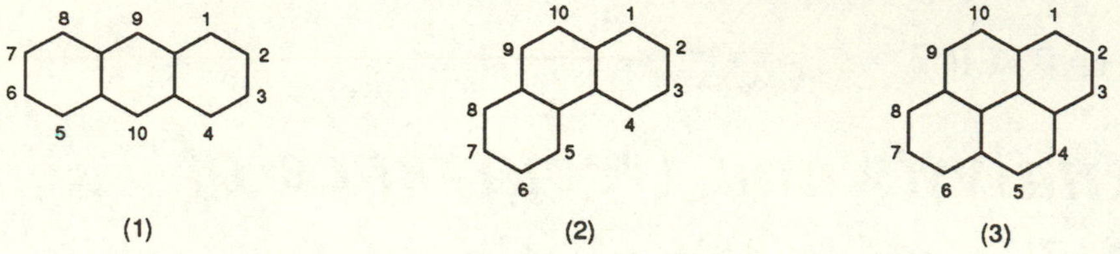

Figure 1

Table 1 *Number of possible PAHs with n six-membered rings*[a]

n	Cata-annellated	Peri-condensed	Total
1	1	0	1
2	1	0	1
3	2	1	3
4	5	2	7
5	12	10	22
6	37	45	82
7	123	210	333

[a] From M. Zander, in 'Handbook of Polycyclic Aromatic Hydrocarbons', ed. A. Bjørseth, Vol. 1. Marcel Dekker, New York, 1983, p. 3.

Not all PAHs show biological activity, and there is often significant variation between isomers. Table 2 compares the carcinogenic activity[7] of selected PAHs. Modern testing tends towards the use of bacterial mutation (Ames test) rather than the induction of tumours in animals; virtually all carcinogenic compounds are mutagenic, although not all mutagenic compounds have been demonstrated to be carcinogenic. Minor constituents of PAH mixtures, such as alkyl derivatives, may make large contributions to the carcinogenic activity of the total mixture. For example, Table 3 shows how certain methylchrysenes, particularly the 5-isomer – one of the most carcinogenic compounds known[8] – may dominate the activity of a given fraction.[9,10]

[7] International Agency for Research on Cancer, 'IARC Monographs on the Evaluation of the Carcinogenic Risk of Chemicals to Humans, Polynuclear Aromatic Hydrocarbons Part I', IARC, Lyon, France, 1983.
[8] S. S. Hecht, M. Loy, R. Mazzarese, and D. Hoffman, in 'Polycyclic Aromatic Hydrocarbons and Cancer', ed. H. V. Gelboin and P.O.P. Ts'o, Academic Press, New York, 1978, Vol. 1, p. 119.
[9] S. S. Hecht, W. E. Bondinell, and D. Hoffman, *J. Nat. Cancer Inst.*, 1974, **53**, 1121.
[10] N. A. Goeckner and W. M. Griest, *Sci. Total Environ.*, 1977, **8**, 187.

Table 2 *Evaluation[a] of the carcinogenic activity of selected PAHs*

Compound	Evidence of carcinogenicity in experimental animals	Evidence from short term tests	Mutagenicity to Salmonella typhimurium (Ames Test)
Fluoranthene	None	Limited	Positive
Pyrene	None	Limited	Positive
Benz[a]anthracene	Limited	Sufficient	Positive
Chrysene	Limited	Limited	Positive
Benzo[b]fluoranthene	Sufficient	Inadequate	Positive
Benzo[a]pyrene	Sufficient	Sufficient	Positive
Benzo[ghi]perylene	Inadequate	Inadequate	Positive
Dibenz[a,h]anthracene	Limited	Sufficient	Positive

[a] From A. Bjørseth and G. Becher, 'Polycyclic Aromatic Hydrocarbons in Workplace Atmospheres: Occurrence and Determination', CRC Press, Boca Raton, Florida, 1986, p. 3.

Table 3 *Comparison of the contents of chrysene and its methyl derivatives in environmental samples[a]*

	Relative Carcinogenicity	Cigarette smoke, ng per cigarette	Coal liquid p.p.m.
Chrysene	+	37	98
1-Methylchrysene	0/+	3	ND
2-Methylchrysene	++	1	102
3-Methylchrysene	++	6	106
4-Methylchrysene	++	0	
5-Methylchrysene	+++	0.5	19
6-Methylchrysene	++	7	64

[a] From M. L. Lee, M. V. Novotny, and K. D. Bartle, 'Analytical Chemistry of Polycyclic Aromatic Compounds', Academic Press, New York, 1981, p. 356.

In recent years significant progress has been made in the understanding of the biological action of PAHs.[4,11] These compounds enter the body by inhalation and ingestion and are distributed to various organs after adsorption by phospholipids and/or complexation by serum albuminates. Enzymatic oxidation by mixed function oxidases (MFO), often termed aryl hydrocarbon hydroxylases, which are most abundant in the liver, followed by hydrolysis to dihydrodiols by aryl epoxide hydrolysases, then occurs. The products are the true active species, the 'ultimate carcinogens', the so-called 'bay region' dihydrodiol epoxides. These compounds form covalent adducts with proteins and nucleic acids. The DNA adducts are thought to initiate cell mutation and eventual malignancy (Figure 2).

[11] A. Bjørseth and G. Becher 'Polycyclic Aromatic Hydrocarbons in Workplace Atmospheres: Occurrence and Determination', CRC Press, Boca Raton, Florida, 1986, Chapter 7, p. 103.

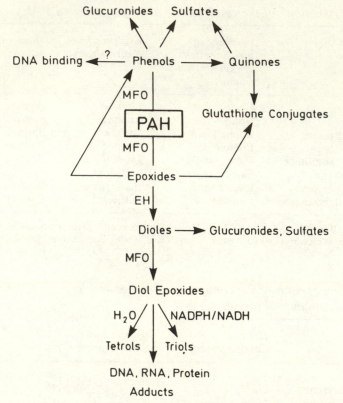

Figure 2 *Enzymatic pathways involved in the activation and detoxification of PAHs: MFO, multifunctional mono-oxygenase, EH, epoxide hydrolase*
(Reproduced with permission from A. Bjørseth and G. Becher 'Polycyclic Aromatic Hydrocarbons in Workplace Atmospheres: Occurrence and Determination', CRC Press, Boca Raton, Florida, 1986)

The formation of PAH-derivative–DNA adducts has been demonstrated by a variety of immunochemical[12] and spectroscopic techniques[4] and followed by HPLC making use of fluorescence.[13] Other reactions involved in PAH metabolism include conjugation[14] of oxygenates (*e.g.* phenols) to glucuronic sulphates and to give water-soluble compounds which are excreted; 1-hydroxypyrene has been shown[15] to be a major metabolite present in urine.

[12] R. M. Santella, L. L. Hsieh, C. D. Lin, S. Viet, and I. B. Weinstein *Environ. Health Perspect.*, 1985, **62**, 95.
[13] R. O. Rahn, S. S. Chang, J. M. Holland, and L. R. Shugart, *Biochem. Biophys. Res. Commun.*, 1982, **109**, 262.
[14] N. Nemoto, in 'Polycyclic Aromatic Hydrocarbons and Cancer', ed. H. V. Gelboin and P. O. P. Ts'o, Academic Press, New York, 1981, Vol. 3, p. 213.
[15] S. D. Keimig, K. W. Kirby, D. P. Morgan, J. F. Keiser, and T. D. Huberg, *Xenobiotica*, 1983, **13**, 415.

2 Origin of PAHs in Food

PAHs in food originate mainly in: the preparation, particularly smoking, and cooking process; by deposition on to vegetable matter of environmental pollutants; or concentration from water pollutant PAH. The relative importance of these sources is discussed in Section 4.

Formation of PAHs is favoured[16,17] by pyrolysis or air-deficient combustion of organic matter at temperatures in the range 500–900 °C and especially above 700 °C. Since peptides, lipids, carbohydrates, terpenes, and leaf pigments have all been shown[2] to be precursors, it is not surprising that cooking at high temperatures under conditions in which pyrolysis products remain in contact with the food leads to the presence of PAHs. The greatest concentrations of PAHs have been shown to arise from the pyrolysis of lipids, especially sterols.[17]

Since incomplete combustion of wood generates PAHs in large quantities,[18] smoked food products are also inevitably contaminated with PAHs. Less avoidable are the PAHs which originate in the environment. There is strong evidence that the origin of PAHs in many foods of vegetable origin lies in atmospheric deposition directly on to the above-ground plant,[19] and that uptake from the soil is of minor importance. Whenever fossil fuels, waste, or vegetation are burned PAHs are released and dispersed into the atmosphere, from which they are deposited directly on to plants.

In definitive work, Jones *et al.* analysed[20,21] grass and wheat grain stored from agricultural experiments conducted on the same site since the 1840s for 16 PAHs. The values for wheat grain harvested between 1979 and 1986 were closely similar to those reported for cereals[22] by Dennis *et al.* in 1983. The patterns of PAHs detected were similar over the extended time series studied, in spite of changing patterns of fuel use, apparently because of the influence of biodegradation, photo-oxidation, and vaporization during transport and retention of PAHs in the environment. A very marked decline in PAHs in grass samples since the 1920s, with corresponding, though lesser, declines for wheat grain parallels the improvement in UK air quality over the same period.

The origin of large amounts of PAHs in certain foods of marine origin is less thoroughly studied. It seems likely, however, that this is most likely to

[16] J. P. Longwell in 'Nineteenth Symposium (International) on Combustion', The Combustion Institute, 1982, p. 1339.

[17] I. Schmeltz and D. Hoffman, in 'Carcinogenesis – A Comprehensive Survey', ed. R. I. Freidenthal and P. W. Jones, Raven Press, New York, 1976, Vol. 1, p. 225.

[18] D. G. DeAngelis, D. S. Ruffin, J. A. Peters, and R. B. Reznik, 'Source Assessment: Residential Combustion of Wood', EPA-600/2-80-0426, Research Triangle Park, North Carolina, 1980.

[19] G. Grimmer and D. Duvel, *Z. Naturforsch.*, 1970, **256**, 1171.

[20] K. C. Jones, J. A. Stratford, K. S. Waterhouse, E. T. Furlong, W. Giger, R. A. Hites, C. Schaffner, and A. E. Johnston, *Environ. Sci. Technol.*, 1989, **23**, 95.

[21] K. C. Jones, G. Grimmer, J. Jacob, and A. E. Johnston, *Sci. Total Environ.*, 1989, **78**, 117.

[22] M. J. Dennis, R. C. Massey, D. J. McWeeny, M. E. Knowles, and D. Watson, *Food Chem. Toxicol.*, 1983, **21**, 569.

lie in the concentration by shellfish of traces of pollutant PAHs in sea water. Levels two or three orders of magnitude above background have been reported[23] in lobsters from areas receiving discharges from coal and coking plants.

3   Analysis of PAHs in Food

Individual PAHs are generally present in foods at the p.p.b. ($\mu g\,kg^{-1}$) level. Three steps are therefore most often involved in analysis: extraction and clean-up, chromatographic separation, and spectroscopic identification.

3.1   Extraction and Clean-up

Extraction of PAHs from oils and fats is straightforward if the materials are soluble in a solvent such as cyclohexane. Recoveries near 100% for PAHs added at the 2 and 10 p.p.b. level have been demonstrated.[24,25] PAHs may be obtained from fruit and vegetable matter by Soxhlet extraction or sonication with a suitable solvent.[26] For insoluble fats and protein-rich foods such as meat, fish, and cheese, alkaline saponification to release PAHs bound to food components is necessary;[27] the samples are then extracted into a hydrocarbon solvent. Thus, only about 30% of PAH is extractable from smoked herring without saponification, which increases the extraction yield to almost 100%.[25]

The food extracts so obtained inevitably contain substantial quantities of interfering materials other than PAHs. Enrichment and clean-up are usually carried out by solvent partition followed by column chromatography. Extraction into dimethylformamide[28] followed by precipitation by water addition and back extraction has been widely applied.[25] Further clean-up on silica gel and separation of PAH fractions on Sephadex LH-20 gel is recommended (Figure 3).[25,28]

3.2   Analysis of the PAH Fraction

The PAH fraction isolated from food is a complex mixture of numerous individual compounds: 146 different PAHs have been identified in automobile emissions,[29] and *ca.* 70 in air particulates.[30,31] The highest resolution

[23] C. J. Musial and J. F. Uthe, *J. Assoc. Off. Anal. Chem.*, 1986, **69**, 462.
[24] J. W. Howard, E. W. Turricchi, R. H. White, and T. Fazio, *J. Assoc. Off. Anal. Chem.*, 1966, **49**, 1237.
[25] G. Grimmer and H. Böhnke, *J. Assoc. Off. Anal. Chem.*, 1975, **58**, 725.
[26] J. T. Coates, A. W. Elzerman, and A. W. Garrison, *J. Assoc. Off. Anal. Chem.*, 1986, **69**, 110.
[27] J. W. Howard, R. T. Teague, R. H. White, and B. E. Fry, jr., *J. Assoc. Off. Anal. Chem.*, 1966, **49**, 595.
[28] M. Novotny, M. L. Lee, and K. D. Bartle, *J. Chromatogr. Sci.*, 1974, **12**, 606.
[29] G. Grimmer, H. Böhnke, and A. Glaser, *Zentralbl. Bakteriol. Parasitenkd. Infektionskr. Hyg. Abt.*, 1977, **164**, 218.
[30] M. L. Lee, M. Novotny, and K. D. Bartle, *Anal. Chem.*, 1976, **48**, 1566.
[31] A. Bjørseth and G. Eklund, *Anal. Chim. Acta*, 1979, **105**, 199.

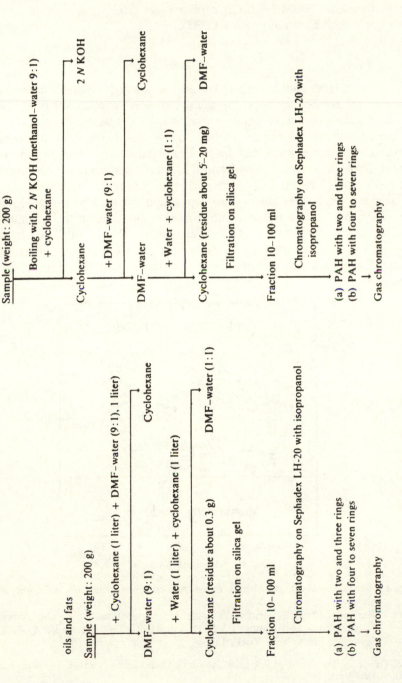

Figure 3 *Schemes for extraction and concentration of PAHs*
(Reproduced with permission from G. Grimmer and H. Böhnke, *J. Assoc. Offic. Anal. Chem.*, 1975, **58**, 725)

possible is therefore required,[28] followed by spectroscopic identification; the methods most generally applied in PAH analysis are thin-layer (TLC), high performance liquid (HPLC), capillary gas (CGC), and most recently, supercritical fluid chromatography (SFC).

3.2.1 TLC

TLC is rapid and simple to apply, and separated components may be recovered for further analysis; the main difficulty is oxidation of PAHs on the active stationary phase. Among available TLC systems,[32] cellulose with 1:1 dimethylformamide/water mobile phase gives the widest range of retention, and isomer pairs may be further separated by changing the water content. Two-dimensional dual band TLC has been shown[33] to give very high resolution of PAHs (*e.g.* Figure 4): the TLC plate consists of alumina and acetylated cellulose layers, and separation occurs first on the alumina (n-hexane/ether mobile phase), with further separation on the second layer with methanol–ether–water. Quantification of PAHs in TLC may be achieved by *in situ* scanning fluorescence densitometry;[32] alternatively separated spots may be isolated from the plate by sublimation or solvent extraction.

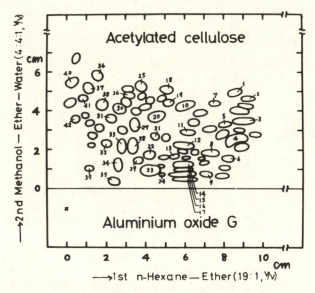

Figure 4 *Two-dimensional dual-band thin-layer chromatogram of pollutant PAH mixture*
(Reproduced with permission from H. Matsushita 'Polycyclic Aromatic Hydrocarbons and Cancer', Vol. 1, ed. H. V. Gelboin and P.O.P. Ts'o, Academic Press, New York, 1978)

[32] J. M. Daisey, in 'Handbook of Polycyclic Aromatic Hydrocarbons', ed. A. Bjørseth, Marcel Dekker, New York, 1983, Vol. 1, p. 397.
[33] H. Matsushita, in 'Polycyclic Aromatic Hydrocarbons and Cancer', ed. H. V. Gelboin and P. O. P. Ts'o, Academic Press, New York, 1978, Vol. 1, p. 71.

3.2.2 HPLC

Several advantages are afforded by HPLC in the analysis of PAHs: selectivity, sensitive and selective detection, and fractionation before the application of other chromatographic or spectroscopic techniques.[34-36]

Selectivity in HPLC is conferred by interactions of solute with both stationary and mobile phases, while UV absorption and fluorescence spectroscopy afford excellent detection methods for PAHs. However, HPLC does not approach the separation efficiency of capillary GC.

Reverse phase separation of PAHs on chemically bonded alkyl silica stationary phases (*e.g.* C_{18}) is the most popular HPLC mode and is preferred over normal phase HPLC. The mobile phase is usually a mixture of a polar solvent and water, the composition of which can be varied in gradient elution. Polymeric phases are most effective,[35] and give separations (Figure 5) in which the shape of the PAH molecule plays an important role.[35,36] Alkyl group number and position also influence retention in reverse phase HPLC. Figure 6 shows representative reverse phase HPLC chromatograms of the PAH fractions from a number of smoked foods.[37]

3.2.3 Capillary GC

The greatest chromatographic resolution available is achieved by capillary GC; the availability of robust fused silica columns, usually 10–25 m long and 0.2–0.3 mm internal diameter, with well deactivated surfaces and cross-linked thermostable stationary phases, now makes this the method of choice for PAH analysis.[36,38,39] Cross-linked polysiloxanes such as SE-54 (methylpolysiloxane with 5% phenyl groups) are most usually employed as stationary phases, but the recent availability of phases with liquid crystalline substituents has given rise to remarkable separations[36] of isomers on the basis of molecular shape – especially the length-to-breadth ratio.

Although flame ionization is the most commonly applied detector in GC of PAHs because of its excellent linearity, sensitivity, and reliability, a range of more sensitive and selective detectors may be used: the electron capture (ECD) and photoionization detectors both exhibit[39] marked selectivity for PAHs. Capillary GC with mass spectrometric detection (GC/MS) represents[40] a powerful combination for PAH analysis. Monitoring of the

[34] S. A. Wise, in 'Handbook of Polycyclic Aromatic Hydrocarbons', ed. A. Bjørseth, Marcel Dekker, New York, 1983, Vol. 1, Chapter 5, p. 183.

[35] S. A. Wise, in 'Handbook of Polycyclic Aromatic Hydrocarbons', ed. A. Bjørseth and T. Ramdahl, Marcel Dekker, New York, 1985, Vol. 2, Chapter 5, p. 113.

[36] J. C. Fetzer, in 'Chemical Analysis of Polycyclic Aromatic Compounds', ed. T. Vo-Dinh, John Wiley, New York, 1989, Chapter 3, p. 59.

[37] F. L. Joe, J. Salemme, and T. Fazio, *J. Assoc. Off. Anal. Chem.*, 1984, **67**, 1076.

[38] M. L. Lee and B. W. Wright, *J. Chromatogr.*, 1980, **184**, 235.

[39] K. D. Bartle, in 'Handbook of Polycyclic Aromatic Hydrocarbons', ed. A. Bjørseth and T. Ramdahl, Marcel Dekker, New York, 1985, Vol. 2, Chapter 9, p. 193.

[40] M. L. Lee, F. J. Yang, and K. D. Bartle, 'Open Tubular Column Gas Chromatography', John Wiley, New York, 1984.

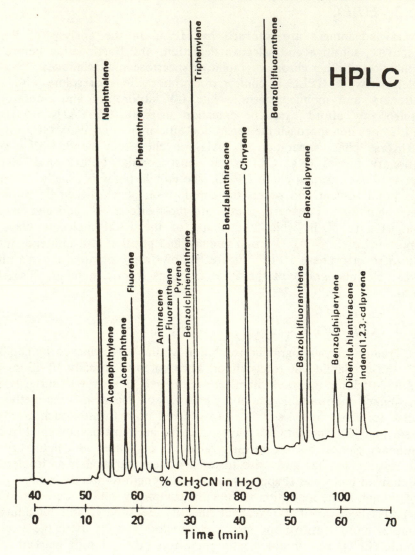

Figure 5 *Reverse phase HPLC separation of priority pollutant PAHs with detection by UV absorption*
(Reproduced with permission from S. A. Wise, W. J. Bonnett, and W. E. May, in Polycyclic Aromatic Hydrocarbons: Chemistry and Biological Effects, ed. A. Bjørseth and M. J. Dennis, Battelle Press, Columbus, Ohio, 1980)

total ion current yields a 'universal' detection chromatogram; alternatively single-ion monitoring allows focusing on one *m/z* value characteristic of the group of compounds under study, so that only the target isomers give peaks.

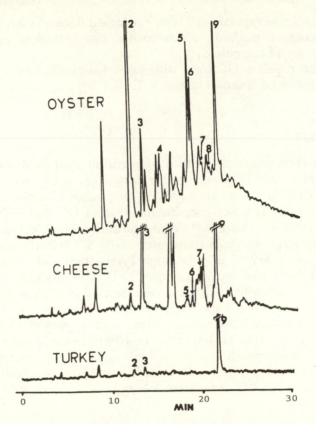

Figure 6 *HPLC chromatograms for extracts of smoked foods with fluorescence detection: 2, fluoranthene; 3, pyrene; 5, benzo[b]fluoranthene; 7, benzo[a]pyrene; 9, benzo[b]chrysene (internal standard)*
(Reproduced with permission from F. L. Joe, J. Salemme, and T. Fazio, *J. Assoc. Offic. Anal. Chem.*, 1984, **67**, 1076)

Selected ion monitoring at *m/z* 202, 228, 252, 278, and 302 detects[41] pyrene and fluoranthenes, chrysene and its isomers, mono- and di-benzo-pyrenes and fluoranthenes, and the dibenzanthracene isomers. Final identi-fication is from the full mass spectrum of a given peak – molecular weight from the molecular ion, and structural information from the fragmentation pattern, usually by means of a computer search of libraries of spectra.

Once compounds have been identified, further analyses may be made by means of retention indices – systems of reproducible retention parameters. For PAHs, a system based on the internal standards naphthalene, phenan-threne, chrysene, and picene is particularly useful. Values of the indices of 310 PAHs and their analogues have been published.[42]

[41] J. F. Lawrence and J. F. Webster, *J. Agric. Food. Chem.*, 1984, **32**, 789.
[42] D. L. Vassilaros, R. C. Kong, D. W. Later, and M. L. Lee, *J. Chromatogr.*, 1982, **252**, 1.

Quantitation in capillary GC is best achieved if sample injection is by the cold on-column method, which avoids discrimination against higher molecular weight compounds.[40]

A typical capillary GC trace with flame ionization detection of a PAH fraction from food is shown in Figure 7.

3.2.4 SFC

Although chromatography with a supercritical fluid as mobile phase was reported more than 20 years ago, the advantages of SFC have only recently been realized: SFC allows analysis of compounds of limited volatility and thermal stability at lower temperatures than in GC, but with high resolution on capillary columns.[44] While most PAHs detected in food are sufficiently stable for high-temperature GC, SFC clearly has applications for PAHs with MW > 300; the large PAHs from carbon black and coal tar[45] have been analysed by capillary SFC. The products of atmospheric oxidation or metabolic transformation discussed in Section 5 clearly represent an area of potential application of SFC.

Use of a supercritical mobile phase with a packed (HPLC) column permits[46] more rapid analysis than in HPLC mode: Figure 8 shows the SFC chromatogram of a coal-derived oil, in which the separation (within 9 minutes) of benzopyrenes and benzofluoranthenes is similar to that obtained by capillary GC of the same mixture.

3.2.5 Spectroscopic Identification

Identification in PAH analysis is also made possible from characteristic UV absorption and fluorescence spectra.[47,48] Fluorimetric detection in HPLC analyses has been recommended and chromatographic peak identities may be confirmed by using UV and fluorescence detectors in series.[37] Alternatively, the entire fluorescence spectrum of a given peak may be scanned and compared with standards[37] (for example, Figure 9). A variety of advanced spectroscopic procedures for PAHs both with and without prior separation

[43] L. Kolarovic and H. Traitler, *J. Chromatogr.*, 1982, **237**, 262.
[44] B. W. Wright and R. D. Smith, in 'Chemical Analysis of Polycyclic Aromatic Compounds', ed. T. Vo-Dinh, John Wiley, New York, 1989, Chapter 4, p. 111.
[45] R. C. Kong, S. M. Fields, W. P. Jackson, and M. L. Lee, *J. Chromatogr.*, 1984, **289**, 105.
[46] I. K. Barker, J. P. Kithinji, K. D. Bartle, A. A. Clifford, M. W. Raynor, G. F. Shilstone, and P. A. Halford-Maw, *Analyst (London)*, 1989, **114**, 41.
[47] E. L. Wehry, in 'Handbook of Polycyclic Aromatic Hydrocarbons', ed. A. Bjorseth, Marcel Dekker, New York, 1983, Vol. 1, p. 323.
[48] M. Zander, in 'Chemical Analysis of Polycyclic Aromatic Compounds', ed. T. Vo-Dinh, John Wiley, New York, 1989, Chapter 6, p. 171.
[49] B. P. Dunn and R. J. Armour, *Anal. Chem.*, 1980, **52**, 2027.
[50] K. Takatsuki, S. Suzuki, N. Sato, and I. Ushizawa, *J. Assoc. Off. Anal. Chem.*, 1985, **68**, 945.

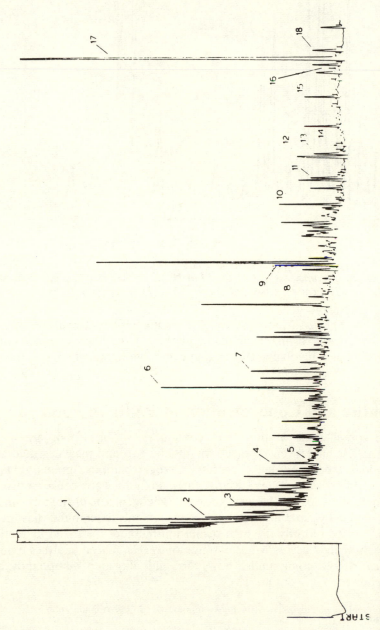

Figure 7 *Capillary GC profile of PAHs from sunflower oil with FID detection. Selected peak identification: 3, fluorene; 4, phenanthrene; 6, fluoranthene; 8, chrysene; 10, benzo[b]fluoranthene; 13, benzo[a]pyrene; 15, indeno[1,2,3-d]pyrene; 17, benzo[b]chrysene (internal standard)*
(Reproduced with permission from L. Kolarovic and H. Traitler, *J. Chromatogr.*, 1982, **237**, 262)

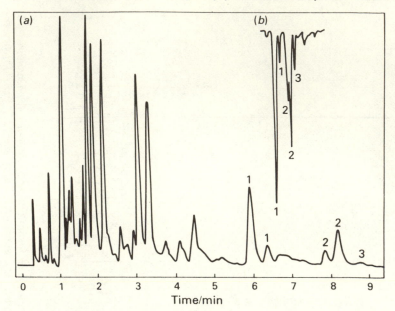

Figure 8 *Chromatograms of coal-derived PAHs, (a) SFC on 25 cm ODS column, CO₂ mobile phase; (b) GC on 14 m capillary column. (1), benzofluoranthenes; (2), benzopyrenes, (3) perylene*

remain to be explored in food analysis:[51] multicomponent analysis may be achieved by line-narrowing methods, and these have been complemented by new techniques involving IR, Raman, and photoionization spectroscopies.

4 Identity and Concentration of PAHs in Food

Extensive compilations of the PAH content of various foods are available.[52-55] PAHs have been reported in smoked fish and meats, grilled and roasted foods, root and leaf vegetables, vegetable oils, grains, plants, fruits, seafood, beverages, *etc.* Many early analyses were concerned only with the determination of the well-known carcinogen benzo[a]pyrene and a recommended TLC method for screening for this compound has been proposed;[56] more recently the high-resolution methods discussed in Section 3 have been applied across a wider range of PAHs. These analyses extend only to *ca.* 10–20 compounds, however, and alkylated compounds are seldom reported.

[51] 'Chemical Analysis of Polycyclic Aromatic Compounds', ed. T. Vo-Dinh, John Wiley, New York, 1989.
[52] J. W. Howard and T. Fazio, *J. Assoc. Off. Anal. Chem.*, 1980, **63**, 1077.
[53] M. T. Lo and E. Sandi, *Residue Rev.*, 1978, **69**, 35.
[54] J. Santodonato, P. Howard, and D. Basu, *J. Environ. Path. Toxicol.*, 1981, **5**, 1.
[55] M. S. Zedeck, *J. Environ. Path. Toxicol.*, 1980, **3**, 537.
[56] G. Grimmer and J. Jacob, *Pure Appl. Chem.*, 1987, **59**, 1735.

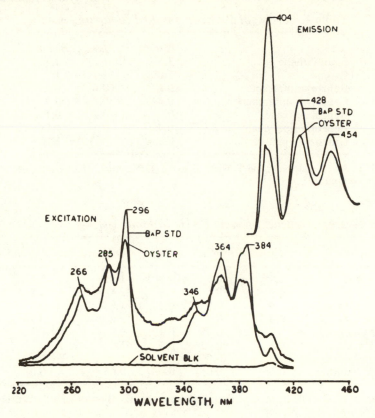

Figure 9 *Excitation and emission spectra of standard benzo[a]pyrene and a component with the same retention time isolated from HPLC analysis of an oyster
extract*
(Reproduced with permission from F. L. Joe, J. Salemme, and T. Fazio,
J. Assoc. Offic. Anal. Chem., 1984, **67**, 1076)

Examples of analytical results for smoked (Table 4) and unsmoked
(Table 5) products and for meat cooked in different ways (Table 6) are
given here. There is general consistency[22,57] among results for similar
sample types, usually at the $1-10 \mu g \, kg^{-1}$ (p.p.b.) level, but occasionally
higher in smoked or grossly contaminated foods; it is nonetheless true that
both intra- and inter-laboratory tests on identical materials have shown
intermediate or poor precision.[58-60] Capillary GC with MS detection has
been suggested[61] as the most reliable procedure because of the superior
resolving power and detection specificity.

[57] M. J. Dennis, R. C. Massey, D. J. McWeeny, and D. H. Watson, in 'Polynuclear Aromatic
Hydrocarbons', ed. M. W. Cooke and M. J. Dennis, Battelle Press, Columbus, Ohio, 1982,
p. 405.
[58] J. F. Lawrence and D. F. Weber, *J. Agric. Food Chem.*, 1984, **32**, 794.
[59] G. Grimmer and J. Jacob, *Pure Appl. Chem.*, 1987, **59**, 1729.
[60] J. F. Uthe and C. J. Musial, *J. Assoc. Off. Anal. Chem.*, 1988, **71**, 363.
[61] K. D. Bartle, M. L. Lee, and S. A. Wise, *Chem. Soc. Rev.*, 1981, **10**, 113.

Table 4 *Levels[a] of PAHs in smoked food ($\mu g\ kg^{-1}$)*

	Cheese	Ham	Bacon
Fluoranthene	3.5	1.2	1.1
Pyrene	7.0	1.7	1.1
Benz[a]anthracene	0.8	0.6	0.4
Benzo[b]fluoranthene	0.4	0.2	0.2
Benzo[a]pyrene	0.5	0.2	0.1
Benzo[ghi]perylene	0.4	0.1	ND
Dibenz[a,h]anthracene	0.1	ND	ND

[a] From F. L. Joe, J. Salemme, and T. Fazio *J. Assoc. Offic. Anal. Chem.*, 1984, **67**, 1076. ND: not determined.

Table 5 *Average content[a] of some PAHs in foods ($\mu g\ kg^{-1}$)*

	Cereals	Meat	Fish	Oils and fats	Vegetables
Fluoranthene	1.4	0.5	0.8	1.8	1.5
Pyrene	1.9	0.6	0.8	2.8	1.2
Benz[a]anthracene	0.4	0.05	0.1	1.0	0.2
Chrysene	0.8	0.2	0.7	1.2	0.9
Benzo[b]fluoranthene	0.2	0.05	0.1	0.9	0.2
Benzo[a]pyrene	0.3	0.05	0.1	1.6	0.1
Benzo[ghi]perylene	0.3	0.05	0.1	1.3	0.1
Dibenz[a,h]anthracene	0.05	0.01	0.03	0.05	0.01

[a] From M. J. Dennis, R. C. Massey, D. J. McWeeny, M. E. Knowles, and D. Watson, *Food Chem. Toxicol.*, 1983, **21**, 569.

Table 6 *Levels[a] of PAHs in differently cooked meat ($\mu g\ kg^{-1}$)*

	Charcoal-broiled steaks	Barbecued ribs
Fluoranthene	43	49
Pyrene	0.9	1.4
Benz[a]anthracene	1.4	3.6
Chrysene	0.6	2.2
Benzo[a]pyrene	5.8	10.5
Benzo[ghi]perylene	6.7	4.7

[a] From T. Fazio and J. W. Howard, in 'Handbook of Polycyclic Aromatic Hydrocarbons', ed. A. Bjørseth, Vol. 1, Marcel Dekker, New York, 1983, p. 496.

In addition to PAHs, the presence of an enormous number of heterocyclic analogues is possible in food, usually arising from environmental contamination from fossil-fuel sources. These include polycyclic aromatic nitrogen heterocycles and their sulphur and oxygen isosteres (Table 7). Various schemes have been devised for the concentration and separation of hetero-compounds, usually employing column chromatography followed by

Table 7 *Classification of PAHs*

Structure	Name	Class
	Acridine ⎱	Polycyclic aromatic nitrogen heterocycles (PANH)
	Carbazole ⎰	
NH₂	2-Aminoanthracene	Amino polycyclic aromatic Hydrocarbons (APAH)
NO₂	2-Nitroanthracene	Nitro polycyclic aromatic hydrocarbons (NPAH)
CN	2-Cyanoanthracene	Cyano polycyclic aromatic hydrocarbons (CPAH)
	Dibenzothiophene	Polycyclic aromatic sulphur heterocycles (PASH)
	Dibenzofuran	Polycyclic aromatic oxygen heterocycles (PAOH)

capillary GC with selective detection[62,63] (flame photometry for PASHs and thermionic detection for PANHs). Although gross contamination with PASHs of fish from a coal-product polluted river has been demonstrated by such methods,[63] a simpler analytical scheme based on HPLC with fluorescence detection suggested[64] that PANHs were not present at levels ≥5 p.p.b. in salmon.

The formation of highly carcinogenic NPAHs in smoked foods by reaction of PAHs with nitrogen oxides during the smoking process has

[62] D. W. Later, M. L. Lee, K. D. Bartle, R. C. Kong, and D. L. Vassilaros, *Anal. Chem.*, 1981, **53**, 1612.
[63] D. L. Vassilaros, P. W. Stoker, G. M. Booth, and M. L. Lee, *Anal. Chem.*, 1982, **54**, 106.
[64] F. L. Joe, J. Salemme, and T. Fazio, *J. Assoc. Off. Anal. Chem.*, 1986, **69**, 218.

been postulated;[65] alternatively, fruit and vegetables originating from sites exposed to road traffic could also be contaminated with the NPAHs known to be present in diesel emissions.[66] In fact, only three of 27 samples analysed contained detectable levels of NPAHs, and these were all foods consumed indirectly.

5   Significance of Contamination of Food by PAHs

While there is little firm epidemiological evidence for the risk of PAHs from food, dietary intake has been identified as the principal route for exposure to PAHs for non-smokers. 'Normal' diets contain[22,67,68] significant concentrations of a variety of PAHs, as is clearly shown by Table 5. Some smoked foods, contaminated shellfish, *etc.* contain larger concentrations, but these are less often consumed. In fact, the concentrations in Table 5, when multiplied by the weight of the food in the average diet, lead[22] to amounts which are of the same order of magnitude as those in the smoke from *ca.* 20 high-tar cigarettes[69] and even more low-tar cigarettes (Table 8).

Although it can be argued that the ingestion routes are insufficiently similar to permit comparisons, ingestion of PAH is a complex process – a significant proportion of inhaled PAHs are removed from the lung into the

Table 8 *Daily intake (μg) of PAHs from food and tobacco smoke*

	From food and beverages[a]	From 20 cigarettes[b]
Fluoranthene	1.0	1.7
Pyrene	1.1	1.4
Benz[*a*]anthracene	0.2	0.5
Chrysene	0.5	0.1
Benzo[*b*]fluoranthene	0.2	0.1
Benzo[*a*]pyrene	0.3	0.3
Benzo[*ghi*]perylene	0.2	0.1
Dibenz[*a,h*]anthracene	0.03	0.1

[a] From M. J. Dennis, R. C. Massey, D. J. McWeeny, M. E. Knowles, and D. Watson, *Food Chem. Toxicol.*, 1983, **21**, 569.
[b] From M. L. Lee, M. Novotny, and K. D. Bartle, *Anal. Chem.*, 1976, **48**, 405.

[65] M. J. Dennis, G. C. Cripps, I. Venn, R. C. Massey, D. J. McWeeny, and M. E. Knowles, in 'Polycyclic Aromatic Hydrocarbons', ed. A. Bjørseth and M. J. Dennis, Battelle Press, Columbus, Ohio, 1983, p. 229.
[66] R. Nakagawa, *Mutat. Res.*, 1983, **124**, 201.
[67] J. P. Tuominen, H. S. Pysala, and M. Sauri, *J. Agric. Food. Chem.*, 1988, **36**, 118.
[68] L. Shuker and B. G. Bennet, 'Exposure Commitment Assessments of Environmental Pollutants', Monitoring and Assessment Research Centre, University of London, Technical Report, 1988, Vol. 6.
[69] M. L. Lee, M. Novotny, and K. D. Bartle, *Anal. Chem.*, 1976, **48**, 405.

gastrointestinal tract by mucociliary clearance and swallowing, so that the doses in Table 8 may well be comparable. These involuntary dose levels from generally consumed foods would appear to be high enough to cause disquiet, and should cause at least similar concern to that over the perceived health hazards of diesel particulate emissions,[70] in which PAHs are strongly implicated. The major origin of dietary PAHs in the collection of combustion-generated pollutants by grain fields is not proven, but clearly merits further investigation. Increasing industrial development worldwide is likely to lead to large increases in anthropogenic emissions and further accumulation of PAHs.

Whereas the relatively high concentrations of PAHs in smoked and barbecued foods have, in the past, led to the targeting of such materials as health risks, it now seems more likely that cereals,[22,67] especially in the form of flour, are a greater hazard.

PAHs are thermally stable products of numerous combustion reactions, but they are nonetheless chemically reactive, particularly towards oxidants such as ozone and singlet molecular oxygen, and with hydroxyl radicals and oxides of nitrogen and sulphur.[71] These reactions have important implications for analytical procedures, especially during extraction and clean-up and the exposure of PAHs on TLC plates to ozone; reactions of anthracene, benzo[a]pyrene, and benz[a]anthracene are particularly rapid with half lives of 0.15, 0.58, and 1.35 hours respectively.[72]

Atmospheric reactions of PAHs could also significantly influence the health risks from PAHs in food. Nitration of PAHs to form NPAHs which are potent direct acting mutagens has been demonstrated;[71] such compounds are known atmospheric[73] and engine exhaust[66] constituents and therefore may be expected to be collected on crops. However, as discussed above, Dennis et al. found[65] detectable levels of NPAHs in only a few foods, none of which is consumed directly.

Of greater concern is the so-far uninvestigated possibility that atmospheric oxidation products of PAHs may be similar to those produced by in vivo metabolic activation, and that hence more potent carcinogens could be deposited on crops and find their way into food by routes similar to those followed by the PAH burden. Thus, the highly mutagenic 4,5-epoxide of benzo[a]pyrene has been detected[74] in airborne particulates, and shown to originate by ozonolysis of the parent. Reaction with singlet oxygen yields quinones of anthracene and benzo[a]pyrene, the latter being converted into the mutagenic 1,6-, 3,6-, and 6,12-diones, while direct

[70] M. P. Walsh, S. A. E. Technical Paper Series No. 871072, Washington, DC, 1987.
[71] K. A. Van Cauwenberghe, in 'Handbook of Polycyclic Aromatic Hydrocarbons', ed. A. Bjorseth and T. Ramdahl, Marcel Dekker, New York, 1985, Volume 2, Chapter 10, p. 351.
[72] M. Katz, C. Chan, H. Tosine, and T. Sakuma, in 'Polynuclear Aromatic Hydrocarbons: Third International Symposium on Chemistry and Biology', ed. R. W. Jones and P. Leber, Ann Arbor Science Publishers, Ann Arbor, Michigan, 1979, p. 171.
[73] T. Nielson, B. Seitz, and T. Ramdahl, Atmos. Environ., 1984, 18, 2159.
[74] J. N. Pitts, D. M. Lokensgard, P. S. Ripley, K. A. Van Cauwenberghe, L. Van Vaeck, S. D. Schaffer, A. J. Thill, and W. L. Belser, Science, 1980, 210, 1347.

conversion of chrysene to mutagenic derivatives has also been shown.[75] Many carbonyl derivatives of aromatics are known to play an important role in human biochemistry.

[75] D. A. Lane, in 'Chemical Analysis of Polycyclic Aromatic Compounds', ed. T. Vo-Dinh, John Wiley, New York, 1989, Chapter 2, p. 31.

CHAPTER 4

Food Production Contaminants: Control and Surveillance

JONATHAN R. BELL and DAVID H. WATSON

1 Introduction

The term food production contaminants can be used to describe a wide variety of chemicals that might be present in food, such as residues of pesticides and veterinary drugs, chemicals migrating into food from packaging, certain naturally occurring toxicants, and adventitious chemical contaminants from food manufacture. However, as other chapters in this book describe work on agrochemical residues and food packaging migrants, special emphasis will be given to these groups here.

The processes of control and food chemical surveillance should be closely linked, whether for food production contaminants or the several other groups of chemicals that are found in food. Where both processes are well developed, as in the UK and a number of other developed countries, control of food production contaminants, or indeed any other chemical contaminants in food, draws upon food chemical surveillance as a source of information in a number of ways. These include identifying a control need, providing the necessary scientific detail in support of a control mechanism that is under development, and checking on the effectiveness of controls once in place.

For controls to be effective, therefore, food chemical surveillance must provide a well organized, reliable source of information. In the UK the mechanism that has developed to provide central Government with such information draws on an extensive body of analytical results. Many thousands of samples of food are analysed every year, mainly by chemical techniques, such as gas chromatography, high performance liquid chromatography, and mass spectrometry. This level of technology is essential since the amounts of contaminants present are generally in the region of parts per million or less. For this reason, and because food is a very complex matrix for the analyst, carefully developed methods of extracting the contaminants from food are essential.

In developing such extraction methods it is important for the analyst:

To take full account of the role that the resulting data will play in aiding the control process;
To ensure that the results have been subjected to an adequate quality assurance procedure;
To devise automated procedures wherever possible; and
To consider information that is already available in deciding what work needs to be done.

The UK food chemical surveillance programme is co-ordinated by the Steering Group on Food Surveillance (SGFS),† a senior UK Government advisory committee. The role and *modus operandi* of this Group have been stated as follows:[1]

'1. The Steering Group (on Food Surveillance)'s function is to identify and evaluate problems, and to propose practical solutions taking into account legal and other constraints.
2. Wherever possible the Steering Group should anticipate problems and act to prevent them.
3. Resources should be used in a cost effective fashion.
4. The Steering Group should publish its intended programmes of work and their results.
5. Special attention should be given to the surveillance of foods consumed by specific groups of people whose diet might put them at more risk than the "average" consumer.
6. Surveillance of emergent foods* and of foodstuffs produced by novel methods should be carried out, once their use is established.
7. Working parties reporting to the Steering Group should actively seek close co-operation with those bodies responsible for environmental surveillance, as and when appropriate.
8. The criteria by which the effectiveness of the surveillance programme of the Steering Group on Food Surveillance may be assessed, are as follows:

8.1 The purity and nutritional quality of the UK food supply should be sustained and enhanced. Contamination of food should be reduced to the lowest practicable level. The highest priority should be given to those contaminants that are believed to be harmful. The food intake of the population should be nutritious.

---

[1] Steering Group on Food Surveillance, 'Food Surveillance 1985 to 1988', Food Surveillance Paper No. 24, HMSO, London, 1988.

†Recently renamed the Steering Group on Chemical Aspects of Food Surveillance.
*Foods that were not previously commonly consumed, for example new varieties of vegetables.

8.2 The confidence of the public and scientists in food surveillance should be increasing.'

The SGFS provides guidance to ten working parties which develop and co-ordinate surveillance work in their specific areas (Figure 1). The programmes of these working parties consist largely of surveys and related research, much of which is on the development of analytical methods. The results of this work are not only considered by the SGFS but by the other Government senior advisory committees shown in Figure 1. The interests of these committees cover the full breadth of issues relevant to the work of the SGFS. The Veterinary Products Committee (VPC) advises Ministers on the licensing of veterinary drugs. The Advisory Committee on Pesticides (ACP) advises Ministers on any matter relating to the control of pests, including which pesticides may be approved for use in the UK. Clearly it is important, in this context, that the results of analytical surveys of drug or pesticide residues, as well as the toxicological and other health-related aspects of these surveillance data, are fully explored by these two committees. The Food Advisory Committee, also referred to in Figure 1, provides expert advice on all food safety issues associated with the chemical contamination of food. As this committee has no extensive toxicological expertise of its own it seeks advice on health-related issues from the Department of Health's Committee on Toxicity of Chemicals in Food, Consumer Products and the Environment.

Solid lines between committees indicate formal links whilst broken lines show informal connections.

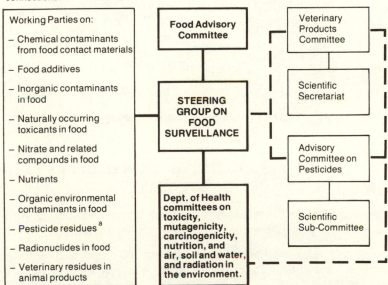

Note: a: This Working Party reports to both the Steering Group on Food Surveillance and the Advisory Committee on Pesticides.

Figure 1 *The Steering Group in relation to other committees*

The considerable body of information, intelligence, and judgement that derives from the food surveillance system is published in a series of reports.[2] Figure 2 shows a selection of these. As well as providing a source of data, and comment and advice from the expert committees on specific issues, these Food Surveillance papers also chart the growth of the programme of work over the past decade or so. Central to this growth has been the development of work on food production contaminants discussed in the latter part of this chapter. However, to set this work in context it is necessary to consider the system of control applied to these substances in the UK.

Figure 2 *Food Surveillance Paper covers*

2 Control of Food Production Contaminants in the UK

Legislative control over residues and food contact materials is mainly derived from the powers contained in three Acts of Parliament:

For *pesticide residues*: control derives from the Food and Environment Protection Act 1985. Part III of this Act provides powers to set maximum levels for pesticides. Specific power to control pesticide residues is provided by Clause 16(2)(k) which says that 'Ministers

[2] Steering Group on Food Surveillance, Food Surveillance Papers Nos. 1 to 30, HMSO, London, 1978–90.

may . . . . by regulations . . . specify how much pesticide or pesticide residue may be left in any crop, food or feeding stuff . . . .'.

For *veterinary drug residues*: the Medicines Act 1968 defines safety, quality, and efficacy as the criteria to be considered in licensing veterinary medicines. In this context safety includes that of consumers as well as those using or receiving the medicines and environmental safety.

For *food contact materials*: control is exercised under the European Communities Act 1972. As this implies, controls derive from European Community legislation to which the UK Government, supported by industry, has made and continues to make a major contribution.

Legislative powers deriving from the first two Acts provide a statutory basis for the advice provided to Government on the licensing of veterinary products and the approval of pesticides by the VPC and ACP respectively. All products must be approved before they can be sold and these Committees consider applications for approval and advise Ministers of their views. In reaching their conclusions both committees take into account the levels of residues likely to remain in the product after treatment. Thus control on residues of pesticides and veterinary medicines in this country is exercised through a pre-approval system for the precursor chemicals. Food chemical surveillance provides an important source of information about the effectiveness of these controls in restricting the resulting levels and incidences of residues in the food supply. It also provides a check on whether the products are being used as approved.

This approach to controlling residues is mirrored in many other countries, such as the USA and Canada, and many of the UK's partners in the European Community. Harmonization of controls on pesticide and veterinary drug residues, and perhaps to a lesser extent on food contact materials, is now becoming a major activity worldwide. Much of the scientific work on the safety of the substances involved is carried out under the aegis of the World Health Organization and the Food and Agriculture Organization, through the respective committees of the Codex Alimentarius Commission. Their published recommendations (for example in reference 3) have an international influence on the control of food production contaminants. They are often taken into account by governments when setting national controls and when agreeing international controls such as those applying within the European Community.

Although legislation provides an important basis for the control of many food production contaminants, it is sometimes desirable to use other methods in place of or in addition to this. The options include:

Collaborative research to resolve problems. A technological solution to a problem can be very effective in the right circumstances. Collaboration,

[3] Joint FAO/WHO Food Standards Programme, Codex Alimentarius Commission, 'Guide to Codex Recommendations Concerning Pesticide Residues, Part 7. Codex Guidelines on Good Practice in Pesticide Residue Analysis', Document CAC/PR7-1984, FAO/WHO, Rome, 1984.

for example between scientific experts involved with control and industrial production, can help to resolve problems which have a high technical content without the need for legislation.

Codes of practice are effective where a responsible industry can overcome a problem by applying rules of good practice.

Advice to consumers by Government that is clear, soundly based, and well communicated can provide an effective means of reducing exposure to chemical contaminants in food, particularly where they are associated with processing of food in the home.

The choice of which approach to follow depends on a variety of factors. But most important of all is the need to ensure consumer protection and choice.

The consumer is the intended beneficiary of any action taken to control the presence of chemicals in food and it is important to inform him or her about the measures that are being implemented in their names. An example of this in the UK is the recent publication of a booklet about pesticides and food. This booklet entitled 'Pesticides and Food: A Balanced View' is available free on request. It explains in straightforward language the basis for pesticide usage and the controls on this and on pesticides residues in the UK. This is one of a series of booklets on food safety issues now available (from Food Sense, London SE99 7TT). The series also includes the booklet 'Food Surveillance' which provides the public with the types of information discussed in this chapter.

3 Surveillance of Food Production Contaminants in the UK

The Government food chemical surveillance system described earlier is used extensively to study migrants from food contact materials and residues of pesticides and veterinary drugs. Although approaches used in each case differ in some details the net result is the same, namely the generation of data on the levels and likely intakes of these from food. Intake data, in particular, provide a common core to much of the surveillance work on chemical contaminants in food.

The value of assessing intakes is illustrated by some recently reported work of the Working Party on Pesticide Residues. This working party used surveillance data to estimate the average intakes of a wide variety of pesticides.[4] The resulting information (Table 1) showed that the intakes were very low and, where comparison with previous information was possible, that they had decreased in some cases. This exercise also allowed comparison with Acceptable Daily Intakes (ADIs) established by national and international scientific bodies. The ADI is a toxicological term which has been defined[4] as 'the amount of a chemical which can be consumed

---

[4] Steering Group on Food Surveillance, 'Report of the Working Party on Pesticide Residues: 1985–88', Food Surveillance Paper No. 25, HMSO, London, 1989.

Table 1 *Computed average intakes and Acceptable Daily Intakes for pesticide residues found in total diet studies in 1981 and 1984–85*

	Intakes (mg per person per day)		
Pesticide	1981	1984–85	Acceptable Daily Intake
Total DDT	0.0020	0.0005	1.4
Dieldrin	0.0007	0.0005	0.007
γ-HCH	0.0020	0.0005	0.7
Chlorpyrifos	NC	0.0001	0.7
Malathion	NC	0.0001	1.4
Pirimiphos-methyl	NC	0.0018	0.7
Cypermethrin	NC	<0.0001	3.5
Permethrin	NC	0.0001	3.5
Quintozene	<0.0004	0.0002	0.49
Tecnazene	0.0029	0.0043	0.7

Note: NC: Not calculated since no residues were detected in the 1981 total diet study.

every day for an individual's entire lifetime in the practical certainty, on the basis of all the known facts, that no harm will result'. This potentially powerful measure provides a quantitative value with which to compare data derived from food chemical surveillance. In the case of the information in Table 1 the comparisons for each pesticide demonstrated that intakes were considerably lower than the respective ADIs. The use of this approach to safety assessment is less widespread for veterinary drug residues and chemicals migrating from food contact materials since fewer ADIs are available. Although there is a growing acceptance of the need to assess these food production contaminants in this way, this is heavily dependent on the development of the necessary body of toxicological information from which ADIs can be derived.

Where ADIs are not available, food chemical surveillance must rely upon other measures of hazard to identify where problems may exist. Sometimes the detection alone of a production contaminant in food identifies a problem. This is the case where presence of the contaminant indicates illegal usage of a pesticide or veterinary drug. The following example, from UK work on veterinary drug residues,[5] illustrates this and also demonstrates the value of transmitting food chemical surveillance information to enforcement authorities. Figure 3 summarizes the surveillance results for residues of a group of synthetic hormone growth promoters – the stilbenes – in samples from cattle, calves, pigs, and sheep in Great Britain from 1981 to 1988. This work was started purely as a surveillance exercise – to quantify the levels and incidences of one stilbene compound, diethylstilboestrol, which was licensed for use as a pig feed

[5] Steering Group on Food Surveillance, 'Anabolic, Anthelmintic and Antimicrobial Agents', Food Surveillance Paper No. 22, HMSO, London, 1987.

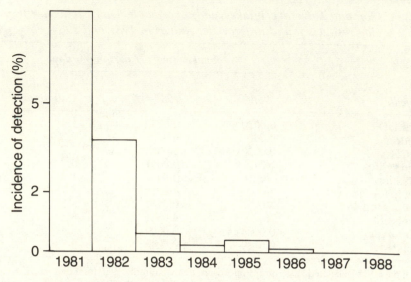

Figure 3 *Stilbene residues. The use of stilbene growth-promoting hormones in food producing animals such as cattle, calves, pigs and sheep was prohibited in 1982. Those found guilty of their illegal use have been fined. As the graph shows, a marked decrease in the incidence of residues has been measured.*

additive. However, fairly soon after the work had been started the use of this and other stilbene compounds in food-producing animals was prohibited in the UK. Although the incidence of residues declined rapidly, some evidence of the continuing use of this hormone was found. Following further investigation, two prosecutions resulted. By 1987 there was no evidence of further usage.

In general, however, food production contaminants are present in food as the result of legal usage and in the great majority of cases the levels found do not indicate a problem. Nevertheless it is prudent to reduce the levels of food production contaminants as far as possible.

Whereas the study of residues of pesticides and veterinary drugs via surveillance is largely based on the analysis of food samples drawn from the general food supply, much of the surveillance work on food contact materials in the UK has involved the study of model systems in the laboratory. In this way interference from similar environmental contaminants to those arising from the materials in question can be taken into account. Although this approach is now being increasingly supplemented with the surveillance methods used in work on residues, there is still value in continuing to use model systems as well. An example of this approach was the recently reported[6] work on migration of the plasticizer DEHA [di-(2-ethylhexyl) adipate], from PVC cling film into test pieces of cheddar cheese. Figure 4 summarizes the kinetics of this migration.

[6] Steering Group on Food Surveillance, 'Survey of Plasticiser Levels in Food Contact Materials and in Foods', Food Surveillance Paper No. 21, HMSO, London, 1987.

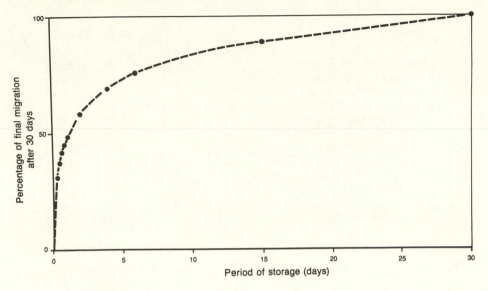

Figure 4 *Migration of DEHA into cheddar cheese at 5 °C*

The surveillance of production contaminants in the food chain is a growing area of science. The numbers of chemicals that could be searched for is large, probably in excess of a thousand. Although there would be little or no point in carrying out surveillance for some of these, such as herbicides that would kill crops rather than leave residues, the remaining list of compounds is still large. Food surveillance, both in the UK and overseas, is probably furthest advanced for pesticide residues. The reasons for this are largely historical as the international scientific community first became interested in pesticide residues in the early 1960s, so that a very extensive database now exists. Many of the lessons learnt in the development of surveillance techniques for pesticide residues have been applied in work on other food production contaminants and indeed other areas of food surveillance.

The development of the UK surveillance programme to give adequate coverage of the large number of chemicals involved and the great variety of foods consumed has necessitated the setting of priorities. An example of this is shown in Figure 5. This is a list of recent projects undertaken by the Working Party on Pesticide Residues. The list ranges from the very specific, such as work on one pesticide, dinoseb, to the very broad such as monitoring of staple items of the diet – bread, milk, and potatoes. These projects were identified as being of high priority based on scientific experience and knowledge obtained from, for example, international fora and the scientific literature.[4] As the potential for pesticide residues to occur can vary according to the commodity, the country of origin, and the type of pest[4] it is essential that care is taken to ensure that the resources available are focused on those areas where residues are most likely to be

Project	Start	Finish
Pesticide residues in eels	1986	1988
Pesticide residues in barley	1987	1987
Pesticide residues in wine	1987	
Monitoring of dietary staples: bread, milk and potatoes	1987	
Residues of dinoseb in UK crops	1987	1989
Pesticide residues in animal feeding stuffs	1987	
Organochlorine residues in venison	1987	1989
Pesticide residues in infant foods	1987	
Trial to determine DDT levels in brassicas	1987	
Monitoring of residues in fruit and vegetables	1988	
Monitoring of residues in cereals and cereal products	1988	
Monitoring of residues in animal products	1988	

Figure 5 *Extract from the programme of the Working Party on Pesticide Residues*

present. The alternative of sampling every type of foodstuff for every possible pesticide residue is impractical and would be very wasteful. Although fewer in number, veterinary drugs should be studied in a similar way for the same reasons. Different factors influence the surveillance of migrants from food contact materials since it is possible to carry out considerable laboratory work on model systems. This in itself can act as a guide to the setting of priorities, as well as contributing data to the overall work programme.[7]

4   Discussion

Objective information about residues *etc.* must provide the key to establishing consumer confidence in the sensible use of chemicals in the production of food and every effort should be made to communicate relevant scientific information both in the form of technical documents intended for the specialist, such as Food Surveillance papers,[2] and in material for a more general audience.

At the present time the degree of hazard arising from chemicals in food is perceived differently by the public and scientists. Figure 6 summarizes the views of some scientists on the ranking of risk from food production contaminants and two other groups of food chemicals. Figure 7 provides a very approximate assessment of public perception of the same chemicals. The duty of scientists is not only vigorously to review data on food production contaminants, but also to communicate their findings as widely as possible. Recipients of such information should include not only the

[7] Steering Group on Food Surveillance, 'Migration of Substances from Food Contact Materials into Food', Food Surveillance Paper No. 26, HMSO, London, 1989.

```
┌─────────────────────────────────────────────────────────────┐
│  Greatest risk                                              │
│                                                             │
│              1.  Natural toxicants                          │
│                                                             │
│              2.  Chemicals migrating from food              │
│                     contact materials                       │
│                                                             │
│                  Pesticide residues                         │
│                                                             │
│                  Drug residues                              │
│                                                             │
│              3.  Food additives                             │
│                                                             │
│  Least risk                                                 │
│                                                             │
└─────────────────────────────────────────────────────────────┘
```

Figure 6 *Some scientists' perception of food chemicals*

```
┌─────────────────────────────────────────────────────────────┐
│  Greatest risk                                              │
│                                                             │
│              1.  Pesticide residues                         │
│                  Food additives                             │
│                                                             │
│              2.  Drug residues                              │
│                                                             │
│              3.  Chemicals migrating from food              │
│                     contact materials                       │
│                                                             │
│              4.  Natural toxicants                          │
│                                                             │
│  Least risk                                                 │
│                                                             │
└─────────────────────────────────────────────────────────────┘
```

Figure 7 *Approximate assessment of public perception of food chemicals*

general public but the media, which in recent times have used information about food production contaminants to raise the public's level of concern about the safety of the foods they eat. It is important to place any hazard in perspective and to direct attention away from substances of little or no risk. In this regard more work needs to be done on assessing the health implications of substances such as naturally occurring toxicants in food about which little is currently known. By quantifying the risk from compounds such as these, the relative risks from dietary exposure to food production contaminants such as drug or pesticide residues or food packaging migrants can be better assessed.

5 Summary

Food production contaminants as a class include a wide variety of chemicals that might be present in food. Three of the more important groups are: pesticide residues, veterinary drug residues, and chemical migrants from food contact materials. The processes of control and surveillance for these chemicals are closely linked. In the UK food surveillance for these and other contaminants is co-ordinated for central Government by the Steering Group on Food Surveillance. The relationships between this interdepartmental committee and other senior advisory committees are described together with their relevant functions.

In the UK, control of food production contaminants is provided through legislation, the co-operation of industry, and government advice to the consumer. Similar means of control are employed in other developed countries, but action on a worldwide basis is essential if contaminants in food are to be properly controlled.

To be effective, surveillance for food production contaminants requires the use of sophisticated analytical methodology, realistic priority setting, and careful interpretation of the resulting data. The use of surveillance data to calculate the intake of chemicals from the diet and the comparison of the results so obtained with Acceptable Daily Intakes, where these exist, has helped to improve assessments of the risk to health posed by these substances.

Scientific work on food production contaminants has an important role to play in informing the public debate about the safety of the food supply. By making the objective information obtained in this way widely available, both specialists and the general public can begin to put the various risks associated with the ingestion of these chemicals into perspective.

CHAPTER 5

Toxicology and Regulatory Control of Components of Food Contact Plastics

RUPERT PURCHASE

1 Introduction

1.1 Food Contact Materials

Although some of the materials traditionally used for packaging and distributing food[1] (paper, cellophane, metals, ceramics, glass, rubber, wood, and fabric) are not without toxicity, the adoption of plastic materials for containing and processing food has presented a major problem for toxicologists, legislators, and analytical chemists for at least three decades.

The stimulus for analytical chemists to identify and quantify the components of plastics (particularly plasticizers) which have migrated into food is discussed in Chapter 6. In this chapter, the legislative problems posed by food contact plastics are described, and some of the toxicological issues arising from their use are discussed.

1.2 Plastics Used for Food Contact Materials

Two types of plastic are used for food contact materials: *thermosets*[2] (plastics which undergo an irreversible change on heating and do not soften[1]), *e.g.* phenol–formaldehyde, urea–formaldehyde, and *thermoplastics*[1-3] (plastics which can be softened by heating and hardened on cooling, repeatedly, provided no chemical decomposition occurs), *e.g.* low density polyethylene – which accounts for the biggest proportion of plastics used in

[1] N. T. Crosby, 'Food Packaging Materials', Applied Science Publishers, Barking, 1981.
[2] F. A. Paine and H. Y. Paine, 'A Handbook of Food Packaging', Leonard Hill (Blackie Group), Glasgow, 1983.
[3] J. H. Briston, in 'Developments in Food Packaging-1', ed. S. J. Palling, Applied Science, London, 1980, Chapter 2.

food packaging, high density polyethylene, polypropylene, poly(vinyl chloride) (PVC), and poly(vinylidene chloride) copolymer (VDC).

High molecular weight polymers themselves have a limited solubility in aqueous and fatty systems and present little toxic hazard. However, the properties of polymers – their resistance to heat and light, flexibility, stability on storage – can be drastically altered by the incorporation of additives.[4]

Concern over the safety-in-use of food contact plastics arises principally from the possible toxicity of low molecular weight and structurally diverse constituents that may be inadvertently or deliberately present in the polymer and leached into food during storage.[1]

Two types of low molecular weight product may be present:[1,4]

(i) Polymerization residues including monomers and oligomers (molecular weight up to 200), polymerization aids (*e.g.* catalysts, cross-linking agents, inhibitors), solvents, emulsifiers and wetting agents, raw material impurities, plant contaminants, decomposition, and side-reaction products.

(ii) Additives and processing aids, *e.g.* antioxidants, antiblocking agents, antistatic agents, heat and light stabilizers, plasticizers, lubricant and slip agents, pigments, fillers, mould release agents, and fungicides.

2 Regulatory Control of the Components of Food Contact Materials

2.1 European Approaches[5]

A number of member states of the European Community (EC) have their own legislation or guidelines based on positive lists of monomers, additives, and other ingredients. These lists may also recommend compositional limits, *i.e.* the maximum amount of each component which may be present in a material. Some countries have also imposed either a specific or a global migration limit on plastic components. For global migration it is generally recommended that plastic materials shall not transfer their constituents to food in levels exceeding either 10 mg dm^{-2} of surface area of the food contact material or 60 mg kg^{-1} of foodstuff (if it is impracticable to estimate the surface area which is in contact with food).

2.2 The UK Approach – The BIBRA/British Plastics Federation Code of Practice

Current regulatory control in the UK requires that materials intended to come into contact with food shall be manufactured such that the components of those materials are not transferred to foods in quantities which could endanger human health or bring about a deterioration in the quality

[4] 'Plastics Additives Handbook', ed. R. Gächter and H. Müller, Hanser Publishers, Munich, 1987.

[5] R. Leimgruber, in ref. 4, chapter 18, p.685.

of the foodstuffs [Statutory Instrument (SI) 1987 No. 1523]. Hitherto the UK has not introduced a strict listing approach to its legislation on the control of plastic components,[6] although there is UK legislation controlling vinyl chloride residues in food contact materials, and the migration of vinyl chloride into food (SI 1987 No. 1523). Instead informal controls have been encouraged through the adoption of a Code of Practice.[7] This document, prepared by the British Plastics Federation (BPF) in co-operation with the British Industrial Biological Research Association (BIBRA), was first issued in 1969 and the latest revision appeared in 1988. The Code of Practice contains the following sections:

- Specifications for the composition of sixteen polymers used for food contact applications
- Criteria for the selection of colorants for food contact plastics
- A list of recommended polymer additives with their maximum level of use in the final product.

BIBRA approval for the inclusion of a polymer additive or ingredient in the Code of Practice is based on the available toxicity and migration data for that compound.

The following information is sought:

- Acute toxicity (single dose or multiple doses over 24 hours)
- Short-term toxicity (multiple doses over periods of up to a few months)
- Long-term toxicity (multiple doses over periods of up to the lifespan of the species investigated)
- Effects on reproduction
- Metabolism (including studies on absorption, excretion, tissue distribution, and metabolic route)
- Sensitization potential
- Genotoxicity.

2.3 EC Legislation on Plastic Monomers and Additives

An EC Working Party on Materials and Articles Intended to Come in Contact with Foodstuffs has had the task of reconciling the differing national views on food contact legislation,[6,8,9] for example the adoption of migration limits into foodstuffs *versus* compositional limits for ingredients in the finished product.

The general strategy of the EC is as follows:[9,10,11]

[6] J. D. McGuinness, *Food Add. Contam.*, 1988, **5** (Supplement No. 1), 525.
[7] Plastics for Food Contact Applications. A Code of Practice for Safety In Use. The British Plastics Federation, London with the co-operation of BIBRA. Revised Edition 1986 (including 1988 update).
[8] L. Rossi, *Food Add. Contam.*, 1988, **5** (Supplement No. 1), 543.
[9] L. Rossi, *Food Add. Contam.*, 1988, **5**, 21.
[10] P. A. Tice, *Food Add. Contam.* 1988, **5** (Supplement No.l), 373.
[11] P. A. Tice and J. D. McGuinness, *Food Add. Contam.*, 1987, **4**, 267.

The ratification of an approved list of substances authorized to be used in the manufacture of food contact plastics
An overall migration limit into food for these substances
Specific migration limits for some compounds
A system of checking migration based on standard testing conditions
Compositional limits for some components in the finished material.

References 10 and 11 give more details of the implications of imposing migration limits, and in particular of the UK PIRA Migration Project which was set up to develop analytical methods for monomers, and provide the required migration data.

In order to draw up positive lists of monomers and additives, an EC toxicological advisory committee, the Scientific Committee for Food (SCF), has been evaluating the available toxicity data on the compounds in current use.[8]

In the SCF reports both monomers and additives have been divided into two Annexes: Annex I-substances for which the SCF was able to express an opinion, and Annex II-substances for which there were insufficient data for the SCF to express an opinion. Each Annex was further divided into various lists (Table 1): Lists 0–4 are substances accepted for use, List 5 are banned substances, and Lists 6–9 are compounds for which there are inadequate toxicity data to enable an opinion to be given, or for which the descriptions are inadequate.

Table 1 *Classification of monomers and additives according to the Scientific Committee for Food (SCF)[a]*

Annex I – Substances for which the SCF was able to express an opinion

List 0
Substances which may be used in the production of plastic materials and articles, *e.g.* food ingredients and certain substances known from the intermediate metabolism in man and for which an ADI need not be established for this purpose.

List 1
Substances for which an ADI, or an equivalent classification has been established by the SCF or by JECFA (Joint FAO/WHO Expert Committee on Food Additives).

List 2
Substances for which a TDI or t-TDI has been established by SCF.

List 3
Substances for which an ADI or TDI could not be established, but where the present use could be accepted. Some of these substances are self-limiting because of their organoleptic properties, or are volatile and therefore unlikely to be present in the finished product. For other substances with very low migration, a TDI has not been set but the maximum level to be used in any packaging material is stated. This is because the available toxicological data would give a TDI very much higher than the maximum likely intakes arising from present uses of the monomer or additive.

Table 1 *(cont.)*

List 4 (for additives)
Substances for which an ADI or TDI could not be established, but which could be used if the substance migrating into food or food simulants is not detectable by an agreed sensitive method.

List 4 (for monomers):
  List 4A (for monomers)
  As list 4 for additives (see above)
  List 4B (for monomers)
  Substances for which an ADI or TDI could not be established, but which could be used if the levels of monomers residues in materials and articles intended to come into contact with foodstuffs are reduced, as much as possible.

List 5
Substances which should not be used.

Annex II - Insufficient data for the SCF to express an opinion

List 6
Substances for which there exists suspicion about their toxicity and for which data are lacking or are insufficient.

  List 6A
  Substances suspected to have carcinogenic properties. These should not be detectable in food or in food simulants by an appropriate sensitive method for each substance.
  List 6B
  Substances suspected to have toxic properties (other than carcinogenic). Restrictions may be indicated.

List 7
Substances for which some toxicological data exist, but for which an ADI or TDI could not be established. The additional specified information should be furnished.

List 8
Substances for which no or only scanty and inadequate data were available.

List 9
Substances and groups of substances which could not be evaluated due to lack of specificity. These groups should be replaced by individual substances actually in use.

ADI = acceptable daily intake; TDI = tolerable daily intake;
t-TDI = temporary tolerable daily intake.
[a] Taken from EEC Document III/3141/89-EN (Rev.3) (CS/PM/500) Brussels 25 July 1990;
  (*BIBRA Bulletin*, 1990, **29**, 277).

As a result of all this activity a Directive containing lists of permitted monomers was issued in 1990 (90/128/EEC). A draft list of approved additives has also been prepared and should be finalised in the near future.[12] EC Directive 90/128/EEC (23rd February 1990 as corrected on 13th December 1990) divides monomers and other starting substances into

---

[12] *BIBRA Bulletin*, 1990, **29**, 277; EEC Document III/3141/89-EN (Rev.3) (CS/PM/500), Brussels, 25 July 1990.

two categories: Section A, the so called 'Community List', which contains authorized monomers (155 compounds drawn from SCF lists 0–4) and Section B, the 'Optional National List' (368 compounds drawn from SCF lists 6–9). Entries in Section B may continue to be used pending a decision on their inclusion in Section A and while the toxicity and descriptive data requested by the SCF are being collated, but after January 1st 1993 only substances in Section A will be allowed for the manufacture of plastics as food contact materials.

For monomers and additives presently assigned to Lists 6 and 7, the SCF hopes to see toxicological and migration tests completed according to the following timetable so that acceptable daily intakes can be established:[12]

Hydrolysis, migration, mutagenicity studies: 31st December 1990 (monomers), 31st December 1991 (additives); (hydrolysis studies are intended to show that a compound can be hydrolysed *in vitro* to products of either known toxicity, or to intermediate metabolites in man)
28 day and 90 day feeding studies: 31st December 1991 (monomers), 31st December 1992 (additives)
Reproductive and teratology studies: 31st December 1992 (monomers), 31st December 1993 (additives)
Long-term studies: 31st December 1995 (monomers), 31st December 1996 (additives).

Less toxicity data are required by the SCF if there is analytical evidence that a compound has a low specific migration (EEC Document III/3568/89-Final). These requirements are explained in Table 2.

Table 2 *SCF toxicity requirements where specific migration data are available*[a]

Migration data (p.p.m.)	Toxicological tests requested	Decision
0–0.050	Three mutagenicity tests	If positive not to be used; if negative $R^* = 0.050$ p.p.m.
0.050–5 and no suspicion of, for example, carcinogenicity or reproductive toxicity	90 day oral test Three mutagenicity tests Test of biological bioaccumulation	Dependent on toxicological results
5–60 or migration data not presented	Full 'core' set of toxicological tests[b]	Dependent on toxicological results

R^* = restriction recommended by SCF and which can be expressed as specific migration limit or as a mg kg^{-1} body weight or in another manner.
[a] Taken from a Note for guidance of applicants for presentation of a request for assessment of a substance to be used in plastic materials and articles intended to come into contact with foodstuffs. EEC Document III/3568/89 – Final (Brussels 25/7/1990).
[b] Mutagenicity studies; 90 day and long-term toxicity studies; reproductive and teratogenicity studies; absorption, distribution, metabolic, and excretion studies.

2.4 Future Developments

EC legislation on food contact plastics has yet to become law in the UK, though it is expected that the EC monomers directive 90/128/EEC and EC directives on migration testing (82/711/EEC) and recommended food simulants for migration tests (85/572/EEC) will be enacted in UK law shortly (by means of a Statutory Instrument under the 1990 UK Food Safety Act).

With the harmonization of national and EC legislation, the BPF/BIBRA Code of Practice will play a less influential role in advising UK manufacturers and users of plastics. It will continue, however, to recommend ingredients and materials which are not yet covered by EC directives: for example, colorants, processing aids, and the composition of laminates not exclusively composed of plastics.

2.5 Informal UK Controls: Surveillance of Plasticizers in Food

The work of the UK Steering Group on Food Surveillance was mentioned in Chapter 4. Of the thirty 'Food Surveillance Papers' so far published by this group, seven have been concerned with the components of food contact materials. Four of these seven reports have dealt with monomers,[13] and of the last three reports two have summarized analytical data on plasticizer levels in food and also have reviewed the literature on the toxicology of these plasticizers.[14a,14b]

The major plasticizers currently used in food packaging materials are di-(2-ethylhexyl) adipate (DEHA) and polymeric plasticizers (polyesters of dicarboxylic acids and dihydric alcohols, average molecular weight 2000) both of which are used in PVC film, epoxidized soya bean oil (ESBO), acetyl tributyl citrate (ATBC) (used in VDC copolymer film), and di-(2-ethylhexyl) phthalate (DEHP), di-isodecyl phthalate (DIDP), and di-iso-octyl phthalate (DIOP) (used in closure seals for jars).

13a Ministry of Agriculture, Fisheries and Food, Report of the Working Party on Vinyl Chloride. Second Report of the Steering Group on Food Surveillance, HMSO, London, 1978.

13b Ministry of Agriculture, Fisheries and Food, Report of the Working Party on Acrylonitrile and Methacrylonitrile. Sixth Report of the Steering Group on Food Surveillance, HMSO, London, 1982.

13c Ministry of Agriculture, Fisheries and Food, Report of the Working Party on Vinylidene Chloride. Third Report of the Steering Group on Food Surveillance, HMSO, London, 1980.

13d Ministry of Agriculture, Fisheries and Food, Report of the Working Party on Styrene. Eleventh Report of the Steering Group on Food Surveillance, HMSO, London, 1983.

14a Ministry of Agriculture, Fisheries and Food, Report of the Working Party on Chemical Contaminants from Food Contact Materials: Sub-Group on Plasticizers, Survey of Plasticizers Levels in Food Contact Materials and in Foods. Twenty-first Report of the Steering Group on Food Surveillance, HMSO, London, 1987.

14b Ministry of Agriculture, Fisheries and Food, Report of the Working Party of Chemical Contaminants from Food Contact Materials: Sub-Group on Plasticizers, Plasticizers: Continuing Surveillance. Thirtieth Report of the Steering Group on Food Surveillance, HMSO, London, 1990.

The levels of plasticizers which migrate into food wrapped in PVC film or VDC copolymer film are summarized in the most recent Food Surveillance Paper.[14b] The analytical methodology for this work is described in Chapter 6. Foodstuffs were analysed after retail or simulated household use. Maximum daily intakes of each plasticizer were then calculated from the average level of the plasticizer found in each foodstuff, and the average consumption of that food in the UK diet.[14b] These calculations suggest maximum daily intakes of DEHA, ATBC, ESBO, and polymeric plasticizers of 8.2, 1.5, 1.0, and 0.4 mg respectively, though it is recognized that because of assumptions regarding the packaging of a particular foodstuff the true consumption of each plasticizer will be substantially less than these calculations.[14b]

Concern about exposure to plasticizers in the diet gathered momentum when a long-term study carried out in the USA in the early 1980s indicated that DEHP produced liver tumours if administered at high doses in the diet to male and female rats and mice,[15] and that DEHA at high doses produced similar effects in female mice and possibly in male mice.[16]

In their review of the toxicology of DEHA, ATBC, ESBO, and polymeric plasticizers, which accompanied the analytical surveillance of these substances in the diet,[14b] the UK Committee on Toxicity (CoT) noted that there is now a 6000-fold or more margin between the maximum estimated dietary intake of DEHA and the dose which produced carcinogenic effects in mice. The CoT were satisfied with the toxicological profile now available for DEHA (although they did request that a specific genotoxicity test – an intraperitoneal dominant lethal study in rodents – should be carried out), but were critical of gaps in the available toxicity data for the other three plasticizers which they requested should be completed by 1991 or 1992 depending on the test.

Future regulatory control of food contact plastics will increasingly require the packaging industry to supply toxicity data on current and new formulations of plastic ingredients to a defined timetable. In order to secure further development in the range and properties of plastics which can be used to protect food some preliminary assessment of likely toxicity problems would seem prudent.

Computer-aided predictions of toxicological endpoints using structure–activity relationships (SAR)[17] could be a useful tool for the assessment and selection of new polymer additives and monomers. SARs have been successfully used to correlate the short-term hepatic effects of phthalate plasticizers (including those used in food contact materials) and

[15] NTP Technical Report Series 217. Carcinogenesis Bioassay of Di(2-ethylhexyl) phthalate (CAS No. 117-81-7) in F344 Rats and B6C3F1 Mice (Feed Study), NIH Publication No. 82-1773, 1982.

[16] NTP Technical Report Series 212. Carcinogenesis Bioassay of Di(2-ethylhexyl) adipate (CAS No. 103-23-1) in F344 Rats and B6C3F1 Mice (Feed Study), NIH Publication No. 81-1768, 1980.

[17] R. Purchase, J. Phillips, and B. G. Lake, *Food Chem. Toxicol.*, 1990, **28**, 459 and references therein.

their metabolites, with molecular structure. The remainder of this chapter describes the background to this work and the development of SAR for phthalic acid diesters, their metabolites, and some other compounds which produce similar hepatic effects in rodents.

3 SAR and the Components of Food Contact Materials

3.1 SAR and Phthalic Acid Esters

Phthalic acid esters now have a limited use in food contact materials, and the maximum daily intakes of DEHP together with its isomer DIOP from food (bottled beverages) are unlikely to exceed 0.02 mg.[14a] However, the large-scale manufacture of phthalates for use in plastics for other purposes, and the awareness that DEHP is an environmental contaminant, has led to the accumulation of much toxicity data.[18] In particular a considerable effort has been directed towards understanding the mechanisms underlying DEHP's carcinogenic properties in rodent liver, and extrapolating these observations to other species.

DEHP does not appear to be directly genotoxic as it is not mutagenic in the Ames test,[19] nor is there evidence that it or its metabolites bind covalently to hepatic DNA after administration *in vivo*.[20] In short-term feeding studies in rats and mice, DEHP, administered at dietary levels up to 2% w/w, produces liver enlargement, hepatic peroxisome proliferation, and induction of peroxisomal and microsomal fatty acid oxidizing enzymes.[21–23]

Peroxisomes are single membrane-limited cytoplasmic organelles which are common to nearly all eukaryotic cells.[24] Peroxisomes can carry out a diverse range of metabolic functions depending on the cell type, and at least forty enzymes have been found in peroxisomes from different tissues.[25] Irrespective of their origin, all peroxisomes contain at least one hydrogen peroxide (H_2O_2)-generating oxidase, and catalase, an enzyme which converts H_2O_2 into water and oxygen.[22] Amongst other functions, hepatic peroxisomes are capable of oxidizing C_8–C_{22} fatty acids (as their acyl-CoA derivatives) by a β-oxidation pathway.[22] The first enzyme of this pathway, acyl-CoA oxidase, produces H_2O_2, and the cyclic oxidation of a

[18] K. N. Woodward, 'Phthalate Esters: Toxicity and Metabolism', Volumes I and II, CRC Press, Boca Raton, Florida, 1988.

[19] *E.g.* D. Turnbull and J. V. Rodricks, *J. Am. College Toxicol.*, 1985, **4**, 111 and references therein.

[20] *E.g.* A. Von Däniken, W. K. Lutz, R. Jäckh, and C. Schlatter, *Toxicol. Appl. Pharmacol.*, 1984, **73**, 373.

[21] Ref 18. Vol. II, Chapter 1.

[22] J. K. Reddy and N. D. Lalwani, *Crit. Rev. Toxicol.*, 1983, **12**, 1.

[23] E. D. Barber, B. D. Astill, E. J. Moran, B. F. Schneider, T. J. B. Gray, B. G. Lake, and J. G. Evans, *Toxicol. Ind. Health*, 1987, **3**, 7.

[24] P. B. Lazarow and Y. Fujiki, *Ann. Rev. Cell Biol.*, 1985, **1**, 489.

[25] N. E. Tolbert, *Ann. Rev. Biochem.*, 1981, **50**, 133.

single fatty acid molecule will result in the production of several H_2O_2 molecules.[25]

The administration of peroxisome proliferators to rodents produces marked increases in the activities of the enzymes of the hepatic peroxisomal fatty acid β-oxidation cycle.[22,23] Palmitoyl-CoA oxidation in the presence of cyanide ions is used as a specific biochemical marker of this cycle.[26] (Cyanide ions inhibit the competing mitochondrial β-oxidation of fatty acids).[26]

In contrast to the induction of acyl-CoA oxidase and other enzymes of the β-oxidation pathway, and the accompanying production of H_2O_2, H_2O_2-metabolizing enzymes (catalase, cytosolic selenium-dependent glutathione peroxidase) are not markedly induced by peroxisome proliferators.[22]

One explanation for the carcinogenic properties of peroxisome proliferators is that an imbalance in the generation and degradation of H_2O_2 results in increases in the intracellular levels of H_2O_2, leading to the production of reactive oxygen species, oxidative damage to intracellular membranes and/or DNA, and ultimately to the formation of liver tumours.[27,28] However, definitive evidence that this explanation is the primary mechanism responsible for tumour formation is lacking.[29]

In studies on the influence of the alkyl moiety on the peroxisome proliferation properties of phthalates, a series of dialkyl phthalates of varying chain length and branching in the alkyl chain were fed to rats at doses up to 2.5% in the diet for 21 days, and markers of peroxisome proliferation were measured.[23] In the same studies DEHA was also tested, and of the phthalates examined four are used as plasticizers in food contact materials (plastics or regenerated cellulose): DEHP, di-isodecyl phthalate, n-butyl benzyl phthalate, and di-n-butyl phthalate.[14a] The results showed that branched chain esters, namely DEHP, di-isodecyl, and di-isononyl phthalates were more potent peroxisome proliferators than the linear chain phthalates examined (n-butyl benzyl, di-n-butyl, and di-undecyl phthalates). In addition DEHA had a potency comparable with the linear phthalates.[23] Overall there was a ten-fold difference between the weakest and strongest esters in terms of their potency as measured by the induction of cyanide-insensitive palmitoyl-CoA oxidation.[23]

This qualitative relationship between structure and potency was developed in an *in vitro* experiment using five isomeric C_6-phthalic acid monoesters and four isomeric C_8-phthalic acid monoesters.[30] (Phthalic acid diesters are hydrolysed *in vivo* to the monoester and an alcohol; these hydrolysis products and their metabolites are thought to be responsible for

[26] P. B. Lazarow and C. DeDuve, *Proc. Nat. Acad. Sci. USA*, 1976, **73**, 2043.

[27] M. S. Rao and J. K. Reddy, *Carcinogenesis*, 1987, **8**, 631.

[28] J. K. Reddy and M. S. Rao, *Mutation Res.*, 1989, **214**, 63.

[29] B. G. Lake, T. J. B. Gray, A. G. Smith, and J. G. Evans, *Biochem. Soc. Trans.*, 1990, **18**, 94.

[30] B. G. Lake, T. J. B. Gray, D. F. V. Lewis, J. A. Beamand, K. D. Hodder, R. Purchase, and S. D. Gangolli, *Toxicol. Ind. Health*, 1987, **3**, 165.

the hepatic effects of dialkyl phthalates).[21,31]

The linear and branched C_6- and C_8-derivatives included 2-ethylhexyl and cyclohexyl phthalic acid monoesters, whose corresponding diesters are used in food contact materials.[14a]

The monoesters were cultured with rat hepatocytes, and their effects on biochemical markers of peroxisome proliferation, including palmitoyl-CoA oxidation, were determined.[30] All the monoesters produced dose-related increases in enzyme activities, and marked quantitative compound potency differences were also observed. Generally the C_8-isomers were more potent peroxisome proliferators than C_6-isomers, and 2- and 3-ethyl-substituted isomers were more potent than linear and 1-ethyl-substituted analogues. The peroxisomal potency differences found *in vitro* between two branched C_8-isomers (1-ethylhexyl hydrogen phthalate and 2-ethyl-hexyl hydrogen phthalate) were also observed *in vivo*.[30] Oral administration of these two compounds to rats produced dose-related increases in the activity of palmitoyl-CoA oxidation, with the 2-ethylhexyl isomer being more potent than the 1-ethylhexyl derivative.[30] Similarly the *in vitro* differences found between branched (2-ethylhexyl) and linear (n-octyl) hydrogen phthalates were also reproduced *in vivo* by administration to rats of either the diester or monoester derivatives containing these two alkyl substituents.[32] These *in vivo–in vitro* correlations demonstrate the utility of hepatocytes for detecting differences in peroxisome proliferation potencies which have been observed *in vivo*, and for screening compounds which induce hepatic peroxisomal enzymes.

Quantitative structure–activity relationships (QSAR) were developed from the *in vitro* data for phthalate monoesters; hydrophobic, Taft steric, and electronic parameters were calculated for each compound, and these parameters were related to the potency data (induction of palmitoyl-CoA oxidation) using Hansch-related equations.* Comparatively poor correlations were obtained using hydrophobic and steric descriptors, but by using electronic descriptors for the phthalic acid monoesters a good correlation ($r = 0.968$) with the potency data was achieved.[30] The electronic descriptors were based on MINDO/3 molecular orbital calculations; (the overall total nucleophilic superdelocalizability of the molecules and the electron densities at carbon atoms 6 and 18† were used in the QSAR).[30]

This *in vitro* SAR work was extended to a set of known but apparently structurally diverse peroxisome proliferators.[33,34] The compounds chosen

[31] C. Rhodes, T. Soames, M. D. Stonard, M. G. Simpson, A. J. Vernall, and C. R. Elcombe, *Toxicol. Lett.*, 1984, **21**, 103.

[32] B. G. Lake, T. J. B. Gray, and S. D. Gangolli, *Environ. Health Perspect.*, 1986, **67**, 283.

[33] D. F. V. Lewis, B. G. Lake, T. J. B. Gray, and S. D. Gangolli, *Arch. Toxicol., Supplement 11*, 1987, 39.

[34] B. G. Lake, D. F. V. Lewis, and T. J. B. Gray, *Arch. Toxicol., Supplement 12*, 1988, 217.

*$\log_{10}$(biological activity) $= ax_1 + bx_2 + cx_3 + \ldots k$ where x_1, x_2, x_3, *etc.* are structural, hydrophobic, and electronic descriptors, a, b, c, *etc.* are numerical coefficients obtained by regression analysis, and k is a constant.

†Carbon atom 6 is the aromatic ring carbon atom linked to the alkoxycarbonyl group; carbon atom 18 is the atom linked to oxygen in the alkoxycarbonyl part of the phthalate monoester.

included 2-ethylhexyl hydrogen phthalate but also chemicals not related to those used in food contact materials, such as clofibric acid (a hyperlipidaemic drug) and aryloxy-substituted carboxylic acids. Again electronic descriptors based on molecular orbital calculations for these compounds were successfully correlated with peroxisomal proliferation potency data (induction of palmitoyl-CoA oxidation[33] and catalase[34] in primary rat hepatocyte cultures). In addition, similarities were seen from the molecular graphic plots of the overall shape of some of these compounds,[34] suggesting a three-dimensional structural relationship not revealed from their molecular structures.

This SAR work raises the possibility of predicting rodent hepatic peroxisomal effects from calculable molecular and structural properties for compounds which are being considered as ingredients of food contact materials, and whose toxicology will eventually require scrutiny by the SCF.

3.2 Other SAR Techniques

The Hansch approach to QSAR has therefore been successfully applied to specific biochemical markers of the hepatic effects of a single group of plastic additives. However, Hansch relationships are restricted to structurally coherent compounds, acting by a common biological mechanism. SAR techniques which attempt to predict more complex endpoints (*e.g.* carcinogenicity) for a set of structurally disparate compounds are needed for an effective toxicity screening of new monomers and additives.

Of the various SAR methods which have been applied in toxicology,[17,35] a number are based on the association of sub-structural fragments with biological activity. Recent developments in computer software now allow the generation of structural fragments and the prediction of possible metabolites[36] from a given structure. The merger of this software with SAR methods based on sub-structural analysis could be a useful development, and go some way towards reconciling the demands of regulatory authorities with the capabilities of polymer manufacturers and toxicologists for the evaluation of some of the components of food contact materials.

Acknowledgement

I should like to thank my colleagues at BIBRA, and in particular Dr Brian Lake, for their help with the preparation of this chapter.

---

[35] L. Turner, F. Choplin, P. Dugard, J. Hermens, R. Jaeckh, M. Marsmann, and D. Roberts, *Toxicol. in Vitro*, 1987, **1**, 143.

[36] F. Darvas, in 'QSAR in Environmental Toxicology-II', ed. K. L. E. Kaiser, D. Reidel Publishing Co., Dordrecht, Holland, 1987, p. 71.

CHAPTER 6

Contaminants from Food Contact Materials: Analytical Aspects

JOHN GILBERT

1 Introduction

Within the European Community, future regulatory control of food contact materials will increasingly depend on the use of positive lists of permitted substances supported by migration testing.[1,2] The use of a particular monomer or additive in a plastic will therefore be controlled, and the amount that leaches or migrates will also be regulated. For the purposes of operating these statutory controls, and to simplify analysis, most testing will be with simulants rather than with real foods. This strategy was first proposed a number of years ago before recent improvements in analytical techniques. The principal aim of using simulants was to simplify analysis to ensure that enforcement could be carried out in most laboratories and could be carried out with commonly available equipment. There has subsequently been much debate over this approach, particularly as to how well simulants actually do 'simulate' real foods.

In order to assess the validity of these food simulants it has proved necessary in many instances to carry out analysis of foods to determine the extent of migration under actual condition of use. Analysis of foods has also proved necessary for surveillance purposes where information is sought on contamination levels in plastic packaged retail samples or foods prepared in the home in contact with plastics. The aim of much of this surveillance is to assess risk by estimating dietary exposure to a particular food contaminant.

[1] L. Rossi, *Food Add. Contam.*, 1988, **5**, 543.
[2] L. Rossi, *Food Add. Contam.*, 1988, **5**, 21.

The ability of the analyst to carry out determinations of low levels of contaminants in real foods has improved significantly in the past few years. This has been largely because of developments in chromatographic techniques, because of improvements in the specificity of detection methods and through novel applications of chemical approaches to sample preparation. It is these new strategies that are reviewed in this chapter, divided into four areas:

(1) The analysis of volatile species;
(2) The analysis of low molecular weight non-volatile species;
(3) The analysis of high molecular weight polydisperse species that can be degraded and can therefore be analysed as low molecular weight entities;
(4) The analysis of high molecular weight polydisperse species that are non-degradable and have therefore to be analysed intact using alternative methods.

2   Headspace Analysis of Volatile Contaminants

2.1   Static Headspace

The principle of what is termed 'static' headspace analysis is very simple and has been described in a number of specialized texts and reviews.[3-5] The sample is sealed in a vial, heated until equilibrium partition is established between the sample and the gaseous headspace, and then a fixed volume of the headspace gas is withdrawn for analysis. The key word is 'equilibration', which is essential for quantification. The gaseous sample is then analysed by gas chromatography in the normal way (by packed or capillary column, in the latter case by split injection). Quantification is by construction of a calibration curve by spiking into uncontaminated material or by standard addition to the sample.

   The main advantage of headspace analysis is its simplicity – no sample preparation is required except sealing a representative portion of the foodstuff in the vial. The method can be easily automated and the same procedure for analysis of the monomer in a food can also be applied to the analysis of the volatile species in the food contact material itself.

   Limits of detection of $1-15 \mu g \, kg^{-1}$ have been easily achieved for the headspace analysis of various monomers in foods (see Table 1) with the use of flame ionization, nitrogen specific, or electron capture detection as appropriate. In a number of instances there is advantage in using a mass

[3] H. Hackenberg and A. P. Schmidt, 'Gas Chromatographic Headspace Analysis', Heyden, London, New York and Rheine, 1977.
[4] B. Kolb, 'Applied Headspace Gas Chromatography' Heyden, London, New York, Philadelphia, and Rheine, 1980.
[5] B. Kolb, 'Analysis of Food Contaminants', ed. J. Gilbert, Elsevier Applied Science, London, 1984, p.117.

spectrometer as the GC detector, for example for styrene,[6] butadiene,[7] and benzene[8] analysis.

Table 1 *Static Headspace analysis of volatile contaminants in foods*

Contaminant	Detection method	Limit (μg kg^{-1})	Examples	Ref.
Vinyl chloride	FID	1	Cooking oil, orange drink	a
Acrylonitrile	AFID	5	Soft margarine	b
Butadiene	GC–MS (SIM)	1	Soft margarine	c
Styrene	GC–MS (SIM)	1–15	Cream, yoghurt, deserts	d
Vinylidene chloride	ECD	5	Biscuits, potato crisps	e
Benzene	GC–MS (SIM)	1	Cooked meals	f

FID = flame ionization detector; AFID = alkali flame ionization detector; SIM = selected ion monitoring; ECD = electron capture detector.
a J. Gilbert and M. J. Shepherd, *J. Assoc. Publ. Anal.*, 1981, **19**, 39.
b J. Gilbert and J. R. Startin, *Food Chem.*, 1982, **9**, 243.
c J. R. Startin and J. Gilbert, *J. Chromatogr.*, 1984, **294**, 427.
d J. Gilbert and J. R. Startin, *J. Sci. Food Agric.*, 1983, **34**, 647.
e J. Gilbert, M. J. Shepherd, J. R. Startin, and D. J. McWeeny, *J. Chromatogr.*, 1980, **197**, 71.
f S. M. Jickells, C. Crews, L. Castle, and J. Gilbert, *Food Add. Contam.*, 1990, **7**, 197.

The use of the mass spectrometer despite being rather sophisticated as a detection method does offer some particular attractions:

(1) The achievable sensitivity is often better than can be attained with other GC detectors – this means that headspace analysis can be utilized for low volatility compounds with unfavourable partition, for example styrene, where on equilibration the majority of the analyte still remains in solution.
(2) As there is no clean-up prior to headspace analysis, normally reliance is placed entirely on chromatography to ensure separation of the analyte from potential interfering species. Mass spectrometric detection is, however, more specific: the monitoring of a number of chosen ions allows for immediate confirmation and this high confidence in correct assignment of identity enables method development to be carried out rapidly.

Static headspace analysis has been successfully applied to a wide range of food matrices varying from the analysis of, for example, vinyl chloride in foods such as cooking oil stored in PVC bottles[9] to the analysis of benzene in composite foods such as complete microwave meals heated in contact with plastics cookware containing trace levels of benzene contamination.[8]

[6] J. Gilbert and J. R. Startin, *J. Chromatogr.*, 1981, **205**, 434.
[7] J. R. Startin and J. Gilbert, *J. Chromatogr.*, 1984, **294**, 427.
[8] S. M. Jickells, C. Crews, L. Castle, and J. Gilbert, *Food Add. Contam.*, 1990, **7**, 197.
[9] J. Gilbert and M. J. Shepherd, *J. Assoc. Publ. Anal.*, 1981, **19**, 39.

2.2 Dynamic Headspace

In some situations the identity of the volatiles which could be released from the packaging may not be known. In order therefore to characterize these compounds fully (say by mass spectrometry), larger sample sizes are required than can be obtained by the static headspace approach. In these instances 'dynamic headspace' is frequently employed where the headspace gas is exhaustively swept or purged into a trap prior to GC analysis. This can be carried out either employing commercially available equipment or using relatively simple 'home-made'devices.[10,11] This approach has proved useful recently in characterizing volatiles released from microwave susceptor materials. These composites consist of a paperboard substrate, adhesive, aluminium, and PET food contact layer and are used for browning applications in a microwave oven.[12] At the high temperatures produced there are the possibilities not only of volatiles being released from any one of the materials in the composite, but also of charring of the paperboard and crazing of the PET layer. Studies of these materials using dynamic headspace approaches have shown complex chromatogams such as that illustrated in Figure 1 for which there is interest in identification of the compounds prior to assessing whether there is a need for subsequent food analysis.

3 Analysis of Low Molecular Weight Non-volatile Species

Most discrete individual chemical species of low molecular weight that might be present in foods as a result of migration can be determined by procedures involving extraction, sequential clean-up stages usually employing open column chromatography followed by GC or HPLC end-determinations. An initial practical limitation may be the resource requirement for method development, which can prove to be extensive, and a limitation to routine application can be the frequently time-consumimg nature of the clean-up. However, this can be sometimes circumvented by utilizing the high specificity of detection methods such as mass spectrometry and thereby reducing the necessary clean-up. A number of examples from the literature illustrate the analysis in foods of low molecular weight non-volatile species such as antioxidants,[13] stabilizers,[14] and oligomers.[15]

[10] D. L. Heikes, *J. Assoc. Off. Anal. Chem.*, 1987, **70**, 215.
[11] J. Drozd and J. Novak, *J. Chromatogr.*, 1978, **157**, 141.
[12] P. Harrison, *Packaging Technol. Sci.*, 1989, **2**, 5.
[13] R. Goydan, A. D. Schwope, R. C. Reid, and G. Cramer, *Food Add. Contam.*, 1990, **7**, 323.
[14] A. D. Schwope, D. E. Little, D. J. Ehntholt, K. R. Sidman, R. H. Whelan, and P. S. Schwartz, *Deut. Lebensm.-Rund.*, 1986, **82**, 277.
[15] T. H. Begley and H. C. Hollifield, *J. Agric. Food Chem.*, 1990, **38**, 145.

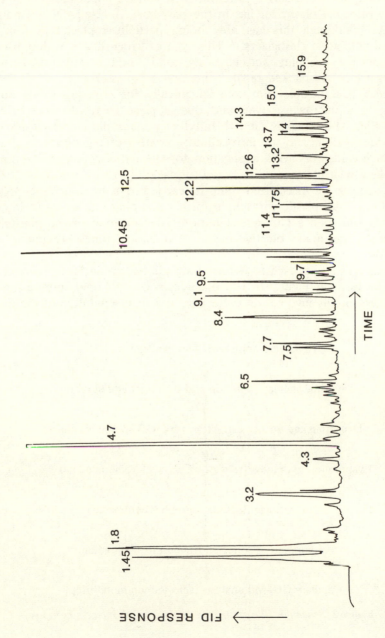

Figure 1 *Capillary GC analysis of volatiles released when susceptor material has been heated for 4 min at 190 °C*

3.1 Plasticizer Analysis

Plasticizers are used to provide flexibility to films and to coatings and a variety of types are used for different food contact applications. For PVC film the principal plasticizer has in the past been di-(2-ethylhexyl) adipate (DEHA)[16] although this has now being partially replaced in films by so-called polymeric plasticizers.[17] This type of 'cling-film' is widely used in the home for wrapping and covering foods, and in supermarkets for covering fresh and cooked meat, fruit, and some vegetables. PVC/PVDC copolymer is a flexible film sold specifically for covering foods during cooking in a microwave oven and this is plasticized with acetyltributyl citrate. Phthalates (mainly dibutyl, butylbenzyl, and dicyclohexyl) are used to provide flexibility to the nitrocellulose coating of regenerated cellulose film which finds very wide application for the packaging of confectionery for cooked meat pies and pasties, and for cakes.[18]

The analysis of plasticized film is carried out by a simple solvent extraction of the plastic with chloroform containing one or more internal standards and then a GC determination. Identification of the plasticizers can usually be made on the basis of retention times compared with standards.

The analytical approach to determining plasticizers in foods is shown in schematic form in Figure 2. The plasticizers are all lipid soluble, so the method essentially involves fat extraction and then separation of the higher

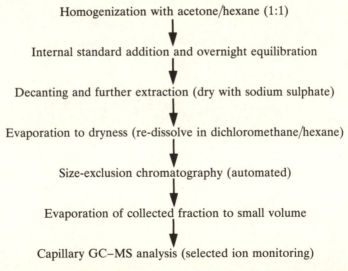

Homogenization with acetone/hexane (1:1)

Internal standard addition and overnight equilibration

Decanting and further extraction (dry with sodium sulphate)

Evaporation to dryness (re-dissolve in dichloromethane/hexane)

Size-exclusion chromatography (automated)

Evaporation of collected fraction to small volume

Capillary GC–MS analysis (selected ion monitoring)

Figure 2 *Schematic illustrating the analysis of plasticizers in foods*

[16] J. R. Startin, M. Sharman, M. D. Rose, I. Parker, A. J. Mercer, L. Castle, and J. Gilbert, *Food Add. Contam.*, 1987, **4**, 385.
[17] L. Castle, A. J. Mercer, and J. Gilbert, *Food Add. Contam.*, 1988, **5**, 277.
[18] L. Castle, A. J. Mercer, J. R. Startin, and J. Gilbert, *Food Add. Contam.*, 1988, **5**, 9.

molecular weight lipids from the lower molecular weight plasticizers by size exclusion chromatography (SEC). The SEC stage of the analysis has the advantage of being amenable to automation. Analysis of the cleaned-up extract is by capillary GC. Although GC analysis using a flame ionization detector (FID) can be effective for some foods, for others, for example confectionery products and some complete cooked meals, interference peaks will be present, obscuring the region of the chromatogram where the plasticizers are expected to elute. In these instances, it is advisable to employ mass spectrometric selected ion monitoring. This has the added advantage that stable isotope-labelled internal standards can then be employed, and the assay becomes one of isotope dilution.[19,20] Although the use of the mass spectrometer does require a high level of sophistication of instrumentation, for large numbers of survey samples, data can be generated quickly, with a high precision and with a high confidence in correct identification of the plasticizers being monitored.

3.2 Analysis of Glycol Softeners

Glycol softeners are used in regenerated cellulose film (RCF) to provide flexibility and are typically used at levels up to 27% w/w in the film. Prior to 1986 monoethylene glycol (MEG) and diethylene glycol (DEG) were widely employed, but have now been replaced by alternative softeners. The softeners now most frequently used are combinations of two or more of the following: propylene glycol, triethylene glycol, poly(ethylene glycol), glycerol, and urea. Although MEG and DEG are still permitted as RCF softeners, recent EC Regulations now restrict the total level of MEG plus DEG in the food as a result of migration not to exceed a limit of $50 \, mg \, kg^{-1}$.

Glycol softeners in RCF can be extracted into acetone or acetonitrile, with the inclusion of butane-1,4-diol as an internal standard, derivatized with silylating reagent, and then analysed by GC. All the glycols can be quantified by this analysis except for poly(ethylene glycol) where only its lower molecular weight components can be detected, although these could be used as a basis for the determination.

The analysis of glycols in foods is particularly difficult because of problems of efficient extraction from the food matrix, the similarity in behaviour of the glycols to other polar molecules present in foods, and the lack of a suitable chromophore or other 'handle' for detection. An effective method has been developed[21] which although rather lengthy can, when applied with care, give good results for the determination of glycols in chocolate and other confectionery products. The method which is shown in schematic form in Figure 3, involves extraction into hot water, defatting

[19] J. R. Startin, I. Parker, M. Sharman, and J. Gilbert, *J. Chromatogr.*, 1987, **387**, 509.
[20] L. Castle, J. Gilbert, S. M. Jickells, and J. W. Gramshaw, *J. Chromatogr.*, 1988, **437**, 281.
[21] L. Castle, H. R. Cloke, J. R. Startin, and J. Gilbert, *J. Assoc. Offic. Anal. Chem.*, 1988, **71**, 499.

with hexane, precipitation of the sugars, TMS derivatization, and finally analysis by capillary GC with an FID. Most problems in this analysis are caused by contamination occurring in solvents, reagents, or from glassware and particular care is necessary in this aspect of sample preparation. Analytical quality assurance checks are therefore essential in the form of frequent analyses of blank samples.

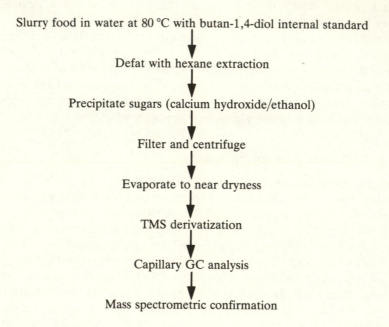

Slurry food in water at 80 °C with butan-1,4-diol internal standard

Defat with hexane extraction

Precipitate sugars (calcium hydroxide/ethanol)

Filter and centrifuge

Evaporate to near dryness

TMS derivatization

Capillary GC analysis

Mass spectrometric confirmation

Figure 3 *Schematic illustrating the analysis of glycol softeners in foods*

3.3   Monomer Analysis

A number of monomers are not sufficiently volatile for headspace analysis, and in some instances where they contain a suitable chromophore HPLC is the appropriate technique to employ. 4-Chlorophenylsulphone is one of the starting substances for the manufacture of polyethersulphone. This monomer can be determined in foods by using a multidimensional HPLC system, where size exclusion and reverse phase chromatography are coupled on-line.[22] In the example shown in Figure 4 the 4-chlorophenylsulphone monomer can be detected at 0.1 mg kg^{-1} in olive oil using HPLC with UV detection, by direct analysis of the oil without a need for extraction or any preliminary clean-up.

[22] R. A. Williams, R. Macrae, and M. J. Shepherd, *J. Chromatogr.*, 1989, **477**, 315.

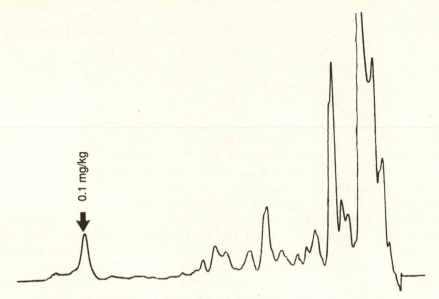

Figure 4 *HPLC analysis of 4-chlorophenyl sulphone monomer spiked at* 0.1 mg kg^{-1} *in olive oil. Automated system employed involving coupled high performance size exclusion chromatography and reverse phase HPLC with UV detection*

4   Analysis of Polydisperse Degradable High Molecular Weight Species

There are many constituents of plastics that are not single species but consist of a distribution of components covering a wide range of molecular weights. Where these polydisperse compounds contain repeat units and are amenable to some form of degradation then the possibility exists of converting to one or more of the constituent parts and then determining these using similar methods to those outlined above. Three examples of plastics additives illustrate this approach.

4.1   Polymeric Plasticizer Analysis

Polymeric plasticizers may be used as partial or complete replacements for conventional plasticizers such as DEHA in PVC cling-film[17]. A typical example of a polymeric plasticizer is a polyester of butane-1,3-diol and adipic acid as shown (1). This plasticizer is in fact a complex mixture of oligomers and has a number average molecular weight of around 1950 – approximately five times greater in relative mass than DEHA.

To determine the migration of polymeric plasticizer, deuterated DEHA

Reoplex 346:
polyester of butane-1,3-diol and adipic acid

(1)

$$D-C-COOC_8H_{17}$$

with structure:

D–C–COOC$_8$H$_{17}$
|
(CH$_2$)$_4$
|
D–C–COOC$_8$H$_{17}$
|
D

(2)

(2) is employed as an internal standard. The approach[23] is based on initial transmethylation of the polymeric plasticizer to form dimethyl adipate (DMA). At the same time the deuterated-DEHA internal standard is also transmethylated to deuterated-DMA, and after automated SEC clean-up, the DMA and [^{2}H$_4$]-DMA are simultaneously measured by selected ion monitoring GC–MS. The limit of detection for the polymeric plasticizer in foods was conservatively set at the lowest level of interest which was 0.1 mg kg^{-1} and the assay was shown to have a precision of about 4%. The method is such that for the analysis of foods containing both DEHA and polymeric plasticizer the assay gives a total level for the two components. If the DEHA is then determined separately the level of polymeric plasticizer in the food can be determined by difference from the total. The method has a high specificity and has been used to analyse a diverse range of foods such as sandwiches, meat, biscuits, cake, and complete microwave cooked meals that have been in contact with plasticized film.

4.2 Epoxidized Soya Bean Oil Analysis

Epoxidized soya bean oil (ESBO) is used as a plasticizer and secondary heat stabilizer in flexible PVC applications at levels of between 3 and 7%, as well as in other plastics where it might, for example, function as a lubricant. The approximate component fatty acid composition of ESBO is 50% diepoxidized, 26% monoepoxidized, and 11% triepoxidized components. An analytical approach to ESBO analysis is possible based on the similar principle of degradation of the triglyceride to its constituent fatty acid components.[24] The approach involving transmethylation under basic conditions converts the triglyceride into its component fatty acid methyl esters. These fatty acid esters do not receive further clean-up from the co-extracted food lipids but are instead derivatized to form 1,3-dioxolanes (3) which are then amenable to direct GC–MS analysis. Quantification is based on an estimation of the diepoxidized fatty acid esters. This assay has a sensitivity and precision similiar to the procedure for monomeric plasticizers and has been applied to a wide range of food types[25] subject to satisfactory blank values being obtained for food samples not having had prior contact with ESBO-plasticized material.

[23] L. Castle, A. J. Mercer, and J. Gilbert, *J. Assoc. Offic. Anal. Chem.*, 1988, **71**, 394.
[24] L. Castle, M. Sharman, and J. Gilbert, *J. Assoc. Offic. Anal. Chem.*, 1988, **71**, 1183.
[25] L. Castle, A. Mayo, and J. Gilbert, *Food Add. Contam.*, 1990, **7**, 29.

$$-(CH_2)_7-CH-CH-(CH_2)_7-$$

(3)

4.3 Analysis of Poly(ethylene terephthalate) Oligomers

Another application of the degradative approach to a single species is the analysis of poly(ethylene terephthalate) (PET) oligomers in foods. PET is being increasingly used both for retail applications such as beverage bottles and for trays and cookware for microwave oven applications. Migration from PET is known to be very low and what little material does migrate probably consists of oligomers of PET, of which the cyclic structures (4) are the most prominent.

The method to measure the total PET oligomer plus terephthalic acid monomer concentrations in foods uses tetradeuterated terephthalic acid as internal standard (5). As with the two schemes outlined above for analysing higher molecular weight analytes, this approach again relies on degradation to a common species.[26] In this instance dimethyl terephthalate is formed both from the complex mixture of oligomers as well as from the deuterated internal standard. Clean-up and GC–MS determination is similar to that described for the other applications. The method has a limit of detection of $0.1 \, \mathrm{mg\,kg^{-1}}$, has a relative standard deviation of 5%, and is applicable to a range of foods from oils and beverages to complete composite meals.

Major cyclic oligomers n = 1-3

(4)

(5)

[26] L. Castle, A. Mayo, C. Crews, and J. Gilbert, *J. Food Protect.*, 1989, **52**, 337.

5   Analysis of Polydisperse Non-degradable High Molecular Weight Species

The final group of compounds which are the most troublesome analytically are those polydisperse species, frequently of relatively high molecular weight, which cannot be broken down into useful smaller portions for analysis. They therefore have to be analysed as a group either by GC, first separating the components and then summing them together to assess the total amount, or by HPLC without separation; both approaches can have disadvantages of poor sensitivity, detection lacking specificity, and poor precision. A recent example that illustrates these problems is the analysis of mineral hydrocarbons in foods.

Mineral hydrocarbons are hydrocarbon fractions derived from petroleum and include mineral oils, liquid paraffins, petroleum jelly, and microcrystalline waxes. They thus encompass a wide molecular weight range and vary from relatively simple mixtures of hydrocarbons that can be resolved chromatographically to complex mixtures that are unresolvable and elute as a broad single component spread over a wide portion of the chromatogram. Food contact applications of mineral hydrocarbons include coatings on cheeses, lubricants for the casings of skinless sausages, and for incorporation in plastics formulations as flow promoters. In this latter application amounts in the range *ca.* 0.5–6.0% of mineral hydrocarbons may be employed, for example in polystyrene cups, tubs, and containers that have a wide range of intended applications in contact with aqueous and fatty foods for normal as well as high temperature applications.

The simplest example of analysis of mineral hydrocarbons has been the study of the migration into cheese from the wax coating. Cheese samples have been extracted into carbon tetrachloride and the hydrocarbon contaminants separated from polar material by a single-stage clean-up using a silica cartridge. The extract was then analysed directly by capillary GC. The lack of complexity of these particular hydrocarbons is illustrated in Figure 5. This means that it is possible to separate individual mineral hydrocarbon components by GC, ensure that there are no interferences from the cheese, and then measure the total mineral hydrocarbon migration by summing the peak areas. In this example C_{16} hydrocarbon was added as an internal standard. The analysis of a core sample of the cheese remote from the wax provided an effective 'blank' sample, and when the chromatogram from this sample was compared with that of a cheese sample obtained from the immediate contact layer with the wax, it was clearly demonstrated that a migration phenomenon was involved.

Most other mineral hydrocarbons that are used in food contact applications have a considerable complexity such that it is not possible to separate the mixtures into individual components. The hydrocarbons thus elute from a GC column as a broad peak which is difficult to quantify, although by the addition of a mixture of internal standards, covering the molecular weight range of interest, it is possible to make some quantitative assess-

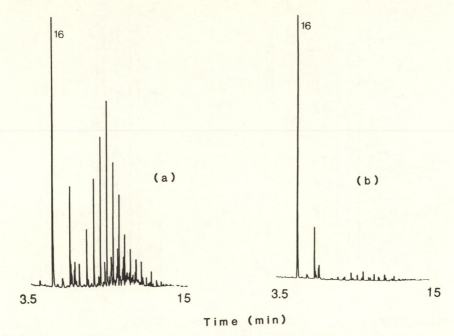

Figure 5 *Capillary GC analysis of mineral hydrocarbon migration from wax coating into cheese. Peak numbered 16 is C_{16} internal standard. (a) Analysis of a sample of cheese cut 1–2 mm from the wax coating; (b) Analysis of a blank sample of cheese obtained from the core of the sample (no contact with wax)*

ment. Figure 6 illustrates the analysis of mineral hydrocarbons in a chocolate drink which has been in contact with a polystyrene vending machine cup containing mineral hydrocarbons. It can be clearly seen from the chromatogram where the drink has been spiked with 0.5 mg kg^{-1} of mineral hydrocarbons that with such a broadly eluting peak there can be no guarantee that food components are not being co-eluted under the peak. Determinations can therefore only be carried out on migration into pre-selected foods which are known to be free of background interferences. The alternative approach of HPLC, although producing a single rather sharper eluting component, lacks sensitivity because of reliance on refractive index detection and still provides no assurance of adequate specificity. Work with these components cannot be carried out on retail samples for surveillance and data can only be generated by setting-up migration experiments.

The best approach to providing reliable migration data into a variety of foods involves incorporating radiolabelled hydrocarbons or fortifying with a mixture of discrete hydrocarbons into the plastics formulation during the fabrication of the containers. This does have the disadvantage that the containers are not commercial products and one is limited to a single

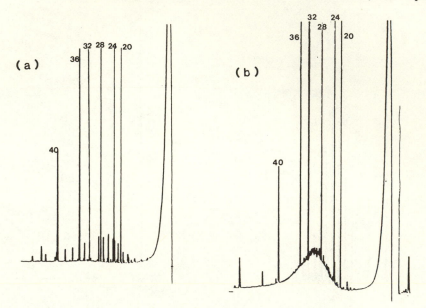

Figure 6 *Capillary GC analysis of mineral hydrocarbon migration from a polystyrene vending machine cup into a hot chocolate drink. Peaks numbered 20–40 are internal standards C_{20} to C_{40}. (a) Chocolate drink containing < 0.5 mg kg^{-1} total mineral hydrocarbon content; (b) Chocolate drink spiked with a total level of 0.5 mg kg^{-1} of mineral hydrocarbons*

formulation. However, a rapid assessment can be made of migration into a diversity of food types and when using radiolabelled compounds there is no pre-requisite that the food should be free of any analytical interferences.

6 Conclusions

This chapter aims to demonstrate that most contaminants that migrate from food contact materials can be analysed at low levels in real foods. The volatile individual species are the easiest to determine because of the possibility of employing the headspace approach. There are, however, strategies emerging for the analysis of the more intractable mixtures that may contaminate foods, although achieving adequate sensitivity and specificity can be a problem. These developments in analytical techniques now enable a far better assessment of migration to be made under realistic conditions of use of food contact materials than has previously been possible.

CHAPTER 7

Use and Regulatory Control of Veterinary Drugs in Food Production

KEVIN N. WOODWARD

1 Introduction

Veterinary medicines, like their human counterparts, are intended to improve or maintain the health of animal species whether these are kept for food production or otherwise. Thus they cover a variety of therapeutic and prophylactic purposes and are administered to food-producing animals, to household pets, and to exotic species. The diseases for which medicines are used range from bacterial infections requiring antibiotic or antimicrobial therapy to external parasitic infestations necessitating topical or immersive treatments with organophosphorus compounds. Many veterinary drugs are administered in an identical manner to human drugs, for example by tablet, injection, or topical application. Others, however, are given in ways which would be unacceptable for human use such as the sheep dips, slow-release anthelmintic formulations, or insecticidal collars for household pets. All medicines used in the UK, whether for veterinary or human use have several things in common. They are all designed to be biologically active, many are potentially toxic, and all are regulated under the Medicines Act 1968 and associated European legislation.

The Medicines Act presents a comprehensive system of licensing for human and veterinary medicinal products, and to qualify for a product licence a medicine must satisfy three criteria: safety, quality, and efficacy. Taking these in reverse order, efficacy is the ability of the drug to accomplish the task that the licence applicant has set for it. Does it work as an antibiotic, as a growth promoter, as an anthelmintic? Quality in this context refers to pharmaceutical quality. The drug must be free from unacceptable contaminants, have the shelf-life and isomeric content claimed for it, and be manufactured to appropriate standards. Safety refers

99

to the inherent ability of a drug product to cause harm, or more specifically not to cause harm to biological systems which become exposed to it. Of course both quality and efficacy have safety implications. An impure drug may contain toxic contaminants or a poorly efficacious medicine may result in a lethal outcome to the disease. Safety on the other hand considers the inherent properties of the medicine. It covers safety to the patient whether human or animal, safety to the operator including veterinarians, farmers, and members of the public, and safety to the environment. In the case of medicines intended for use in food-producing animals, safety to consumers of products containing drug residues is of paramount importance. It is largely this latter aspect which will form the substance of this chapter.

2   Medicines Legislation

Apart from some extremely old statutes governing the quality of medicines, there was no effective control of medicines in the UK prior to 1968.[1] However, following the disastrous sequence of events brought about by the use of the drug thalidomide in the 1960s, the Committee on Safety of Drugs under the Chairmanship of Sir Derek Dunlop was formed, and later drug scrutiny was put on a statutory basis by the passing in Parliament of the Medicines Act 1968.

Similarly, prior to 1968, veterinary drugs had been controlled in the UK by a voluntary scheme aimed at ensuring appropriate safety labelling with the intention of protecting humans, livestock, domestic animals, and wildlife against any potential adverse effects. The Ministry of Agriculture, Fisheries and Food which oversaw the scheme was advised on these aspects by the Advisory Committee on Pesticides and Other Toxic Chemicals and by the Veterinary Products Committee.[1] After 1968, veterinary products were regulated under the Medicines Act.

The Medicines Act provides for a Licensing Authority which has the power to grant – or otherwise – a product licence. The Licensing Authority is defined in the Act as the Health and Agriculture Ministers acting in a joint manner. It is advised on safety, quality, and efficacy matters by expert independent committees established under Section 4 of the Act. For human medicines the Committee on Safety of Medicines provides the advice whilst for veterinary drugs the Veterinary Products Committee (VPC) serves this purpose.

The function of the Licensing Authority is devolved to the Medicines Control Agency (MCA) of the Department of Health for human medicines and to the Veterinary Medicines Directorate (VMD), an Executive Agency and a part of the Food Safety Directorate of the Ministry of Agriculture,

---

[1] M. F. Cuthbert, J. P. Griffin, and W. H. W. Inman, in 'Controlling the Use of Therapeutic Drugs. An International Comparison', ed. W. M. Wardell, American Institute for Public Policy Research, Washington DC, 1978, p. 99.

Fisheries and Food, for veterinary products. Staff in the MCA and VMD provide scientific and administrative secretariats which service the independent expert committees, carry out assessments of safety, quality, and efficacy data, take part in national and international discussions of drug regulation, and issue the product licences and other forms of market authorization provided for under the legislation.

Applications for product licences for veterinary medicines are sent to the VMD which is based on the site of the Central Veterinary Laboratory in Weybridge, Surrey. Specialized assessments of the safety, quality, and efficacy data are carried out by toxicologists, pharmacists, and veterinarians on the staff of the Directorate and reports are prepared which are initially considered by a committee known as the Scientific Secretariat. This is composed of members of the VMD staff and experts drawn from other parts of the Ministry of Agriculture, Fisheries and Food including the Central Veterinary Laboratory and Food Science Division as well as from other government agencies including the Department of Health and the Health and Safety Executive. The Committee has no statutory basis but it serves to give an initial and exceptionally thorough analysis of the data supplied in support of applications, identifying problem areas and shortcomings. The reports, modified where necessary in the light of the Scientific Secretariat's deliberations, are then forwarded for independent expert advice to the VPC.

The simplest outcome to a VPC consideration is to recommend that a product licence be granted. However, factors often prevail which prevent this conclusion. There may be inadequate or insufficient data, more information may be required, or the Committee may decide that the product is not safe, efficacious, or of sufficient quality – or of course a combination of these. Under these circumstances the Committee writes to the applicant under Section 21 of the Medicines Act informing the company concerned that it may not be able to advise the Licensing Authority to grant a licence and listing the reasons for this decision. The applicant is then offered the opportunity to make written representations or to appear before the Committee in order to resolve the problems. This is commonly referred to as the appeals procedure, and if unsuccessful the applicant is given a final opportunity to appeal before the Medicines Commission, a broadly based committee appointed under Section 2 of the Act. The Medicines Commission may uphold or overturn the decisions of the VPC. When the 'appeals' procedure is exhausted the applicant may pursue his case by judicial review.

The Licensing Authority also has powers to deal with existing product licences when adverse effects become apparent or if new data indicate that the product can no longer be regarded as safe, efficacious, or of sufficient quality. For example, a newly published carcinogenicity study may indicate that an active ingredient or excipient has the potential to induce cancer in laboratory animals. In such cases, the Licensing Authority may revoke the product licence, a process which takes some time and during which period the product is allowed to remain on the market. However, if it is believed that the effect is life-threatening, then the licence may be suspended with immediate effect whilst revocation proceedings are simultaneously put into effect. For more minor effects, the Licensing Authority may seek to vary the licence compulsorily so as to include additional warnings on the data sheet or to alter the type of packaging. Licence revocation is subject to the appeals procedures described earlier.

An important part of the licensing system is the Adverse Reaction Scheme. This is similar to the 'yellow card' scheme for human drugs and it allows suspected adverse reactions in animals, or indeed in humans if they become contaminated, to be reported to the Licensing Authority. Consequently, alterations to the data sheet, to the product label or to the licensing conditions can then be made to reduce the incidence of, or avert further adverse reactions.

It is important to note that product licences are not the only market authorizations granted for veterinary drugs under the Medicines Act. The Animal Test Certificate (ATC) is the equivalent of the human Clinical Trials Certificate. It is granted largely for the purpose of establishing the efficacy of a proposed new medicine. As the toxicological and residue profiles are usually poorly established when companies apply for ATCs, carcases of treated food-producing animals are often destroyed to prevent them from entering the human food-chain. Companies may also apply for variations to existing product licences or to ATCs to extend these to other indications or to other species. Again, sufficient data must be supplied to support such applications. A current exercise requiring a considerable amount of input by the VMD Secretariat and the VPC arises from the review of medicines. When the Medicines Act 1968 took effect in 1971, products already on the market in the UK were eligible for Product Licences of Right.[2] These were often granted with only a minimum of data. Although they carry the same status as standard product licences, both the Medicines Act and European legislation required that these products should be reviewed. They are essentially being reviewed as if each was a new licence application and applicants must submit safety, quality, and efficacy data to satisfy the Licensing Authority that the products meet modern standards of assessment. As with new applications, the Licensing Authority is advised on review products by the VPC.

3 European Legislation

In the UK, what might be termed conventional medicines, *i.e.* those given orally, by injection, or topically applied for standard therapeutic reasons *and* those given via the medication of feed, are treated in an identical manner using regulations made under the Medicines Act. This is not the

---

[2] G Jones, 'The Regulation of Medicines in the UK', in 'Pharmaceutical Medicine', ed. D. M. Burley and T. W. Binns, Edward Arnold, London, 1985, p. 149.

case in European Legislation. Conventional medicines are dealt with under the veterinary medicines Directives whereas medicated feeding stuffs are dealt with under a separate Directive. Indeed, each is handled by a separate Directorate General within the European Commission in Brussels. The European legislation governing veterinary medicines will not be discussed at length in this chapter as it as been described in some detail elsewhere.[3] However, the fundamental aspects will be briefly explained in the paragraphs which follow.

The two major Directives which pertain to conventional veterinary medicines in the European Community are Council Directives 81/851/EEC and 81/852/EEC.[4,5] The main objective of the former was to establish the Committee for Veterinary Medicinal Products (CVMP) and of the latter to lay down the regulatory requirements for analytical, toxicological, and pharmacological test guidelines. These have been subsumed into UK law using Medicines Act legislation. The two Directives serve to ensure that regulatory requirements for veterinary medicinal products are the same throughout the Community, whilst the purpose of the CVMP is to determine whether or not a medicinal product complies with the requirements of the Directives. A manufacturer wishing to obtain market authorization in Member States, having already done so in a single State, may follow the so-called CVMP procedure whereby his application is forwarded to that Committee's Secretariat in Brussels and to five other member states for consideration. The opinion of the CVMP will be sought if any of the States raises an objection on safety, quality, or efficacy grounds. If a favourable opinion is given, Member States must then individually decide on granting marketing authorization, and the CVMP is informed accordingly. It is compulsory for all high technology products, namely those derived from recombinant DNA technology or from novel techniques, to be considered in this way.[6]

Medicinal products intended to be added to feed, and other compounds for which no medicinal claims are made, including mineral supplements and amino acids, are considered under the terms of Council Directive 70/524/EEC ('concerning additives in feeding-stuffs') and not under the Medicines Directives.[7] Under article 7.2.A(b) of amending Directive 84/587/EEC, an ingredient is acceptable if 'at the level permitted in

[3] K. N. Woodward, in 'Textbook of Applied and General Toxicology', ed. B. Ballantyne, T. Marrs, and P. Turner, Macmillan, Basingstoke, in press.

[4] Council Directive of 28 September 1981 on the approximation of laws of Member States relating to veterinary medicinal products (81/851/EEC). *Official Journal of the European Communities*, 1981, No. L317, 1.

[5] Council Directive of 28 September 1981 on the approximation of laws of Member States relating to analytical, pharmaco-toxicological and clinical standards and protocols in respect of the testing of veterinary medicinal products (81/852/EEC). *Official Journal of the European Communities*, 1981, No. L317, 16.

[6] Commission of the European Communities' 'The Rules Governing Medicinal Products in the European Community', Volume V, Veterinary Medicinal Products, EC, Luxembourg, 1989.

[7] Council Directive of 23 November 1970 concerning additives in feedingstuffs (70/524/EEC). *Official Journal of the European Communities*, 1970, No. L270, 840.

feeding stuffs, it does not adversely affect human or animal health or the environment nor harm the consumer by altering characteristics of livestock products'. Basically this means that active ingredients to be considered for inclusion into feed must satisfy the same criteria for human, animal, and environmental safety as their more conventional medicinal counterparts.

To gain market authorization within the Community, an active ingredient for inclusion in feed must be entered into either Annex I or Annex II of Directive 70/524/EEC. Entry into Annex I means that the compound in question must be made freely available throughout the EC. Compounds with a less complete data package may enter into Annex II as a transitory measure pending further information. If these data are both forthcoming and satisfactory, the ingredient in question will enter Annex I, but if the data show adverse effects or indeed are not supplied by a sponsor(s) the compound falls from the Annexes and effectively loses market authorization.

To be considered for Annex entry an applicant usually approaches its own national authority, which in the case of the UK is the VMD. Under these circumstances the application is considered like any other product licence application and a VPC opinion is sought. Should this be favourable, officials then guide the application through the EC procedures which take the form of the Standing Committee for Feeding Stuffs and an EC Expert Working Group. Advice on toxicological and other specialized areas is provided by the Scientific Committee on Animal Nutrition or SCAN and ultimately, if all is satisfactory, Annex entry is recommended.

4   Residues

When a drug is given to food-producing animals, the main area of concern is undoubtedly consumer safety arising from the presence of residues. Residues are the small amounts of the drug and its metabolites which remain in animal tissues at the time of slaughter or which are passed into other edible animal products such as eggs, milk, and honey as a result of treatment.

The first stage in assessing the safety of residues is the examination of a package of toxicity data submitted by the licence applicant, so that a toxicological profile of the drug can be established. The information is scrutinized so that a full picture of the biological properties of the compound can be built up and any particular adverse properties can be defined. This process is aided by carefully considering other relevant biological data in addition to information provided by toxicological tests. These will include information on absorption, biotransformation, and excretion in laboratory and target species, data from pharmacodynamic studies, and toxicity data from trials in the proposed target animals. Although the veterinary use of the drug may be new, it may have previously been employed in human medicine and valuable information may be available from this source either in the form of adverse reaction reports or, more rarely, from epidemiological studies.

The simplest case to consider is that of a new molecule solely intended for use in veterinary medicine. An application for this will be accompanied by a series of toxicity reports generated in laboratory animals or in *in vitro* systems and in the target species, accompanied by the results of pharmacological studies, again generated in laboratory and target animals. If the drug is intended for use in food-producing animals, the range of toxicity data required will be extensive and may consist of acute and subacute studies, investigation into effects on reproduction and development, mutagenicity and carcinogenicity studies, and specialist studies to elucidate particular aspects of toxicity, *e.g.* neurotoxicity or teratology. The major goal is to identify a no-effect level from the data package.[8,9] From the no-effect level, an acceptable daily intake (ADI) can be calculated using an appropriate safety factor. The safety factor usually has a value of 100 but for severe forms of toxicity or where there is some degree of uncertainty about the toxicity or the no-effect level, a larger factor may be chosen.[10-12] The ADI, with a knowledge of food intake data, is then used to establish a Maximum Residue Level (MRL). Unfortunately, food intake data are difficult to generate and research is complicated by members of the population with extreme intakes of a particular food commodity such as those noted in ethnic or religious groups. Nevertheless, such studies are conducted, and the Ministry of Agriculture, Fisheries and Food publishes total diet information.[13] In practice, standard food intake data, which include large safety margins, are usually employed.

Once the MRL is established for the active ingredient, withdrawal periods can be determined. The withdrawal period is the time from cessation of dosing until the animal is acceptable for slaughter when the residues of the drug have fallen below the MRL. In order to identify a withdrawal period, groups of animals are given the drug as the intended proprietary formulation using the intended route of administration. It cannot be overemphasized that the ADI and the MRL apply to the active ingredient but the withdrawal period pertains to each of its formulations, as some may be designed to be fast acting and hence be rapidly cleared from the tissues while others may be slow release products and residues

[8] N. H. Booth in 'Veterinary Pharmacology and Therapeutics', 5th Edn., ed. N. H. Booth and L. E. McDonald, Iowa State University/Ames, 1982, p. 1065.

[9] K. N. Woodward, in 'Clinical and Experimental Toxicology of Anticholinesterases', ed. B. Ballantyne and T. Marrs, Butterworth, London, in press.

[10] E. J. Bigwood, *CRC Crit. Rev. Toxicol.*, 1973, **2**, 41.

[11] M. K. Perez, *J. Toxicol Environ. Health*, 1977, **3**, 837.

[12] G. Vettorazzi and B. Radaelli-Benvenuti, 'International Regulatory Aspects for Pesticide Chemicals', Vol. II, Tables and Bibliography, CRC Press, Boca Raton, Florida, 1982, p. 109.

[13] M. E. Peattie, D. H. Buss, D. G. Lindsay, and G. A. Smart, *Food Chem. Toxicol.*, 1983, **21**, 503.

may persist for considerable periods of time. Groups of dosed animals are then slaughtered so that the residue depletion can be monitored, and the time point when levels fall below the MRL identified. The withdrawal period is then specified as part of the licence requirements for the product in question. Eggs, milk, and honey present particular problems. Animals can be retained from slaughter using a suitable withdrawal period until drug residues decay to below the MRL but residues in eggs, milk, and honey do not deplete with time. Consequently, it may be necessary to discard these commodities until levels fall to or below the MRL.

It is of course of very little practical use to expend time and money on toxicological studies and the MRL procedures if withdrawal periods are ignored and residues then enter the food chain at unacceptable levels. There is an obvious need for a policing system. The so-called Residues Directive was published in the Official Journal of the European Communities in 1986 and was incorporated into UK legislation in 1988 in the form of the Animals and Fresh Meat (Examination for Residues) Regulations 1988.[14,15] These regulations make provision for the analysis of tissue samples and for the restriction of animals containing residues above official MRLs. These MRLs have yet to be established; they are currently under consideration by the CVMP's Working Group on Residues. Nevertheless, surveillance for residues in the UK is now extensive and responsibility for it lies with the VMD in collaboration with the State Veterinary Service. In 1989, some 40 000 samples were analysed and this figure is set to increase. As well as monitoring for residues above the MRL, the surveillance system is also a check for the presence of banned substances, including anabolic steroids and thyrostatic compounds. In addition to the residue sampling conducted by the State Veterinary Service National Surveillance Scheme, the Steering Group on Food Surveillance oversees a programme of surveillance through its Working Party on Veterinary Residues in Animal Products. The results of this work are considered by the Department of Health's Committee on Toxicity of Chemicals in Food, Consumer Products, and the Environment and where relevant by its sister Committees on mutagenicity and carcinogenicity, before being put to the VPC. This body then considers the data and any recommendations from the Department of Health's Committees before making its own pronouncements which in problem areas may involve modifying the MRL or the conditions of a product licence. The results and Committee deliberations are published in the form of a Food Surveillance Paper.[16,17]

[14] Council Directive of 16 September 1986 concerning the examination of animals and fresh meat for the presence of residues (86/469/EEC). *Off. J. Eur. Commun.*, 1986, No. L275, 36.

[15] Statutory Instruments No. 848 Agriculture. The Animals and Fresh Meat (Examination for Residues) Regulations 1988, HMSO, London, 1988.

[16] Anabolic, Anthelmintic, and Antimicrobial Agents. The Twenty-Second Report of the Steering Group on Food Surveillance. The Working Party on Veterinary Residues in Animal Products. Food Surveillance Paper No. 22, HMSO, London, 1987.

[17] Food Surveillance 1985 to 1988. Progress Report of the Steering Group on Food Surveillance (1988). The Twenty-Fourth Report of the Steering Group on Food Surveillance, HMSO, London, 1988.

There are some circumstances where an MRL cannot be identified. These present particular problems for both industry and for the regulatory agencies. The best, or perhaps worst example is when the active ingredient of interest is a mutagen or a genotoxic carcinogen.[18,19] In theory no no-effect level can be established for these effects[19] and so an ADI or MRL cannot be calculated. It has been suggested that the no threshold effect and the zero tolerance concepts are untenable,[20] and that various methods can be employed to assess the degree of risk from problem compounds.[21] However, in a world where consumer issues are uppermost in many people's minds, the practice of allowing compounds for which a no-effect level cannot be identified into the food-chain is unacceptable. Drugs containing such compounds are therefore generally reserved for non-food producing animals.

5 Future Prospects

There are numerous pressures which may ultimately shape the require-ments of veterinary medicine licensing, both in the UK and in the European Community as a whole. These include pressures from individual consumers, from the so-called Green Lobby, and from industry. However, perhaps the most fundamental change will be that arising from ideas to create a European licensing agency. At present ideas in this area are perhaps not surprisingly vague but the European Commission has signalled its intentions to establish such an agency.[22,23] It is intended that this organization would bring together the market authorization of veterinary and human medicines in the Community and this route of authorization would become more widely used – as opposed to national licensing – after 1992. So far there are no detailed plans as to the location of the organization but it is hoped that more detailed ideas on its function, siting, and operation will be available in the early 1990s. However, it seems likely that the national licensing operations will need to be in existence for the foreseeable future, to cope with domestic situations and to provide a source of expertise for European bodies.

On the scientific front there are still major issues which need to be overcome in the safety assessment of veterinary drugs. The induction of antibiotic resistance in bacteria has been a contentious issue for some time

[18] W. Flam, in 'Chemical Safety Regulation and Compliance', ed. F. Homburger and J. K. Marquis, S. Karger, London, 1985, p. 1.

[19] F. C. Lu, 'Basic Toxicology, Fundamentals, Target Organs and Risk Assessment', Hemi-sphere Publishing, London, 1985, p. 81.

[20] A. Somogyi, 'Carcinogenic Risks. Strategies for Intervention', IARC Scientific Publications, No. 25, 1979, p. 123.

[21] M. L. Dourson, R. C. Hertzberg, R. Hartung, and K. Blackburn, *J. Toxicol. Ind. Health*, 1985, **1**, 33.

[22] Commission of the European Communities. Memorandum on the Future System for the Authorization of Medicinal Products in the European Community, III/B/6, April, 1989.

[23] Commission of the European Communities. Future System for the Free Movement of Medicinal Products within the European Community (Four Preliminary Draft Proposals), III/3603/90-EN, February, 1990.

with the concern that the use of antibiotic and antimicrobial drugs in veterinary medicine might lead to the emergence of resistant human pathogens. So far there is no firm evidence for this phenomenon.[3]

The recently formed Codex Committee on Residues of Veterinary Drugs in Food (CCRVDF) has begun to consider MRLs for residues of veterinary drugs in edible animal products. It is advised on toxicological and residue matters by the Joint FAO/WHO Expert Committee on Food Additives (JECFA) which has now met three times to consider veterinary medicines. The deliberations of these prestigious bodies are bound to have an effect on both national licensing agencies and on discussions within the European forum. This will be particularly important if non-EC countries adopt Codex MRLs which differ from those chosen by the European Community as this situation would create obvious barriers to commerce and the possibility of export–import disputes would arise. There is an increasing need therefore for harmonization of acceptable residue levels to prevent situations of this type from arising.

It is possible to be certain of only one factor in the future. Consumer and political interest in food quality and safety will ensure that, both domestically and internationally, pressure will continue for stringent safety assessment of ingredients used in veterinary medicines so that the public can be reassured that animal produce is without risk to human health.

CHAPTER 8

Analysis of Veterinary Drug Residues in Edible Animal Products

MARTIN J. SHEPHERD

1 Introduction

Veterinary drugs are used both to cure and to prevent animal infections and infestations. Drugs are also added to animal feeds to improve the efficiency of their conversion into edible tissue. Residues may therefore occur in animal products.

This review focuses on recent applications of physico-chemical methodology and immunoaffinity chromatography to the analysis of drug residues, excluding arsenicals, in edible tissues. Methods for the analysis of veterinary drugs in non-foodstuffs, such as biological fluids, faeces, animal feeds, and pharmaceutical preparations, are not included.

About 100 different drugs are permitted for use in the UK but before any new compound is licensed for farmyard animal use it will have undergone thorough toxicological assessment to ensure that any residues in food do not present a hazard to the consumer.[1] When required, withdrawal periods are stipulated so that the processes of excretion and metabolism of the drug reduce its concentration in animal tissues to safe levels before slaughter. Residues are the subject of extensive government surveillance in the UK[2] and many other countries.

Permissible Maximum Residue Levels (MRLs) for drug residues in animal products are set by a number of organizations including the European Community (EC). MRLs for a given drug will vary depending on species and tissue type. The EC is in the process of establishing mandatory uniform MRLs under Directive 86/469 and the final and

---

[1] K. N. Woodward, Chapter 7 of this volume.
[2] Anon., Anabolic, Anthelmintic and Antimicrobial Agents, Food Surveillance Paper No. 22, HMSO, London, 1987.

provisional limits awaiting adoption are given in Table 1. The structures of these drugs are shown in the relevant part of Section 2. For other drugs, Table 2 lists MRLs used by the UK Veterinary Products Committee to

Table 1 *Provisional and final European Community MRLs*

Compound	Tissue	MRL (μg kg^{-1})	Status	Notes	Structures Figure
Azaperone	K	100	P	a	23
	M, L, F	50	P	a	
Benzylpenicillin	M	50	F		4
	Mk	4	F		
Ampicillin	M	50	F		4
	Mk	4	F		
Amoxicillin	M	50	F		4
	Mk	4	F		
Oxacillin	M	300	F		4
	Mk	30	F		
Cloxacillin	M	300	F		4
	Mk	30	F		
Febantel	L	1000	P	b	2
	M, Mk, K, F	10	P	b	
Fenbendazole	L	1000	P	b	2
	M, Mk, K, F	10	P	b	
Oxfendazole	L	1000	P	c	2
	M, Mk, K, F	10	P	c	
Carazolol	L, K	50	P		23
	M, F	5	P		
Chloramphenicol	M	10	P		6
Dapsone	M, Mk	25	P		24
Ivermectin	L	15	F	d	10
	F	20	F	d	
Levamisole	M, Mk	10	P		24
Nitrofurans group	M, E	5	P	e	12
Dimetridazole	M	10	P	f	15
Ronidazole	M	2	P	f	15
Sulphonamide group	M	100	F		17
	Mk	100	P		
Trimethoprim	M, Mk	50	F		19
Tetracyclines group	K	600	P		20
	L	300	P		
	E	200	P		
	M, Mk	100	P		
	F	10	P		

Tissue: E, egg; F, fat; L, liver; K, kidney; M, meat; Mk, milk.
Status: F, Final; P, Provisional.
[a] Azaperol.
[b] Oxfendazole + oxfendazole sulphone + fenbendazole.
[c] Oxfendazole + oxfendazole sulphone.
[d] B_{1a} component.
[e] All residues with intact 5-nitro structure.
[f] All residues with intact nitroimidazole structure.

determine drug withdrawal periods. Additional MRLs will be set as necessary. Some of these drugs may currently be determined only by microbiological methods. Another EC proposal which will eventually have a considerable impact upon choice of methodology for some laboratories is a draft Directive laying down criteria to be fulfilled by routine and reference methods for the analysis of residues in food-producing animals and their products.

From a regulatory point of view it is important to establish compliance with MRLs. It is also necessary to monitor for possible illegal use of drugs which are either banned or not licensed. Other actions taken to ensure consumer safety include effective surveillance of the food supply to collect information on actual levels of residues. Not all drugs require the same degree of monitoring. Enhanced surveillance may be needed on the basis of toxicological advice, known occurrence of residues following incorrect or

Table 2 *Examples of MRLs used by the United Kingdom Veterinary Products Committee to set withdrawal periods*

Compound	Meat	Milk	Eggs	Structure Figure
		$(\mu g\ kg^{-1})$		
Amphotericin	10			–
Amprolium	500			24
Apramycin	100			–
Bacitracin	$0.7\ IU\ g^{-1}$	$1.2\ IU\ ml^{-1}$	$4.8\ IU\ g^{-1}$	–
Cephalosporins	60	10	30	4
Clopidol	200			24
Decoquinate	1000			–
Dihydrostreptomycin	1000	200	500	1
Erythromycin	300	40	300	11
Ethopabate	500			24
Griseofulvin	200	10		–
Lasalocid	700 (liver)			9
Lincomycin	200			–
Maduramicin	25			–
Monensin	50			9
Nafcillin		20		–
Neomycin	500	150	200	1
Novobiocin	500	150	100	24
Nystatin	7100	1100	4300	–
Oleandomycin	300	150	100	–
Polymixin B	$5\ IU\ g^{-1}$	$2\ IU\ g^{-1}$	$5\ IU\ g^{-1}$	–
Robenidine HCl	100			–
Spectinomycin	500	150	200	–
Spiramycin	25			–
Streptomycin	1000	200	500	–
Tiamulin	400 (liver)			–
Tylosin	200			11
Virginiamycin	100			24

inappropriate use, or to ensure compliance with legislative bans on certain compounds. Some drugs are closely scrutinized because a small number of individuals in the population have the potential for allergic or other adverse reactions to very low concentrations of residues which would have no effect on most consumers. Examples include penicillins and chloramphenicol.

Drug residue analysis is also an important aspect of quality control for companies providing animal-derived foodstuffs. In addition to establishing that raw materials comply with MRLs, other factors may be important. For example, the dairy industry needs to ensure the absence of antibiotics from milk intended for making cheese or other products where a bacterial culture is employed in the manufacturing process.

It has therefore been necessary to develop methods for the determination of drugs at low concentrations in animal tissues, milk, eggs, and honey. This poses special problems for the analytical chemist as animal products are highly complex materials, more so than plant-derived foods, and liable to vary substantially in composition depending upon species, tissue type, and the nutritional regime of the animal. Furthermore, animals give rise to a wider range of metabolites.

In addition to the bulk lipid, protein, and water components, a wide range of low molar mass components are also present. Cattle liver, extracted with acetonitrile and after removal of lipids by size exclusion chromatography, gave a residue of 0.30% by weight of compounds with a molar mass of 400 or less. Simple arithmetic indicates that this could represent 3×10^4 components present at 100 μg kg^{-1} in the liver. Measurement of analytes is often carried out by spectrometry, and quantitation of a single component made in the presence of other co-extractives is obviously susceptible to interference; thus trace level analytical procedures make extensive use of chromatography to purify extracts prior to the actual determination.

It has been estimated[3] that a column efficiency of *ca.* 2×10^5 theoretical plates is required to give a 90% probability that a mixture containing 20 random components will be resolved with unit resolution into individual peaks. Typical HPLC columns and capillary GC columns generate perhaps 10^4 and 10^5 theoretical plates respectively. It is evident therefore that unless the sample extract is very highly purified a single chromatographic separation even at the highest currently attainable efficiency will not by itself be sufficient to eliminate the possibility of interference. However, effective analysis depends as much on selectivity as on efficiency. This may be selectivity of detection or of separation, and these two approaches are typified by mass spectrometry (MS) and immunoaffinity chromatography respectively.

Drug residue analysis is expensive to carry out and methods must be optimized for the purpose required, which will influence both the required

[3] J. M. Davis and J. C. Giddings, *Anal. Chem.*, 1983, **55**, 418.

sensitivity and the extent of confirmation of drug identity. Monitoring a national food supply to provide information on average consumer intakes requires the availability of methods with limits of detection (LoDs) well below MRLs because when drugs are used in accordance with good veterinary practice, residues, if present at all, are likely to occur at very low levels. Full confirmation of positive results is required, while for quality control a simple yes/no answer at the MRL may be all that is required. It is usually the parent drug which is determined but metabolites can be important as for example with nitrofurans and nitroimidazoles.[4]

Stephany has described in detail the quality control aspects of residue analysis.[5] Two important considerations facing any analyst are sample size and tissue type. Sample size will depend initially upon the sensitivity of the detection process used. Nanogram quantities of the drug are likely to be required for each measurement, and in most cases a concentration stage will be required during sample preparation as this represents the residue from 0.01–10 g of tissue. Residues of drugs are often higher in liver and kidney compared with muscle, and these 'target tissues' should be selected for analysis where other constraints permit. Concentrations of oestrogenic hormones for example are typically five times greater in liver than in muscle. Other hormones are lipophilic and for these the target tissue will be fat. Much higher levels of drugs may be present in biological fluids such as bile and urine.

Another factor of lesser significance is sample homogeneity. There are indications that for certain residues, such as dihydrostreptomycin in kidney[6] and benzimidazoles in liver,[7] concentrations may vary within organs. The natural variability of fat distribution in tissue must also be taken into consideration when taking individual samples. The stability of residues in frozen storage varies considerably, although in many cases there are inadequate data. Nitrofurans are degraded within a few days, whereas tetracyclines for example are stable for long periods. If metabolism is rapid it may be undesirable to allow frozen tissue to thaw completely before extraction.

Chemical modification of a parent drug may occur by metabolite formation or its alkylation or acylation of biological macromolecules including proteins[8] and nucleic acids. Some drugs, such as clenbuterol, are tightly but non-covalently held at binding sites on certain proteins and cannot readily be extracted unless the protein is hydrolysed beforehand.

[4] A. Y. H. Lu, P. G. Wislocki, S.-H. L. Chiu, and G. T. Miwa, *Drug Metab. Rev.*, 1987, **18**, 363.

[5] R. W. Stephany, *Belg. J. Food Chem. Biotechnol.*, 1989, **44**, 139.

[6] J. R. Lockyer, A. Bucknell, and I. C. Shaw, in Proc. Euroresidue Conf. Residue Vet. Drugs Food, ed. N. Haagsma, A. Ruiter, and P. B. Czedik-Eysenberg, University of Utrecht, Utrecht, 1990, p. 254.

[7] S. A. Barker, T. McDowell, B. Charkhian, L. C. Hsieh, and C. R. Short, *J. Assoc. Offic. Anal. Chem.*, 1990, **73**, 22.

[8] V. Burgat-Sacaze, A. Rico, and J.-C. Panisset, in 'Drug Residues in Animals', ed. A. G. Rico, Academic Press, New York, 1986, p. 1.

This may be achieved enzymatically, using subtilisin or other proteases. Other drugs, including nitrofurans and nitroimidazoles, undergo extensive metabolism via simple conjugation to polar esters such as sulphates or glucuronides, or more complex chemical modification via oxidation and reduction. The type and extent of metabolism varies between animal species and tissues. Information on metabolite toxicity, metabolic pathways, and protein binding is often unavailable in the published literature.

When presented with a novel sample type it is important to carry out careful tissue spiking experiments to determine the extent to which recovery is affected by protein binding or other factors. Extraction of residues from foods has been reviewed by Petz.[9] The simplest way to evaluate extraction procedures is by incorporation of radiotracers, but few labelled drugs are commercially available. Manufacturers of radioimmunoassay (RIA) kits are a potentially useful source of small amounts of tracer. Drugs containing stable isotopes are ideal internal standards for MS methods and some may eventually be available from the EC Bureau of Reference Materials. However, obtaining even standard samples of drugs is not always straightforward, as many are not available from chemical supply houses and the producer must be approached directly. Addresses are available from trade associations and other sources.[10]

Methods of analysis may be grouped into three classes on the basis of the principle used: physico-chemical, immunochemical, or microbiological. A comparison of these three approaches is given in Table 3. A brief discussion of their advantages and disadvantages follows, with more detailed information on methods for individual drugs given in the section on applications. This review will concentrate on physico-chemical and immunochemical methods.

Table 3 *A comparison of chemical, immunochemical, and microbiological methods for veterinary drug residue analysis in animal tissue*

Method	Sensitivity	Specificity	Confirmation	Sample throughput Single	Batch	Sample prep.	Availability
Chemical	+ to +++	++	+++	+	+	−	+++
Immuno-assay	++ to +++	++	+	+	+ to +++	− to ++	+
Micro-biological	+ to +++	+	−	− to +	+++	++	++

1.1 Physico-chemical Methods

These methods are almost invariably based on chromatographic purification of residues followed by spectroscopic quantitation. They are applicable

[9]M. Petz, *Lebensmittelchem., Lebensmittelqual.*, 1988, **13**, 34.
[10] Anon., APBI Compendium of Data Sheets for Veterinary Products, 1990–91, Datapharm Publications Ltd., London, 1990.

to virtually all drug categories, require relatively short development times, and in particular permit unambiguous confirmation of drug identity by techniques such as MS. Unfortunately, as discussed above, in order to attain adequate specificity it is often necessary to apply protracted sample preparation schemes which severely limit sample throughput. Furthermore, chemical determination of residues at trace levels typical of foodstuffs requires considerable skill in manipulating sample extracts, both to achieve an acceptable recovery of analyte reproducibly and to avoid cross contamination of samples.

Most physico-chemical residue analysis is carried out by extraction of homogenized samples into an appropriate solvent, removal of interferences (clean-up) by liquid–liquid partition, and/or low-resolution chromatography and reverse phase HPLC quantitation of the drug concentration. Commercially supplied prepacked solid-phase extraction (SPE) cartridges have largely replaced laboratory-prepared clean-up columns and instrumentation has improved substantially over the past 20 years, particularly with regard to data storage and reduction. Nevertheless, procedures have yet to take full advantage of recent developments in automated liquid handling, robotics, and selective sample clean-up by immunoaffinity chromatography (IAC) or other specialized chromatography packings.

One innovation has been the development of on-line dialysis–concentration column–HPLC methods. Although early instruments provided only 5–15% transfer of analytes across the membrane, commercial equipment can achieve dialysis efficiencies of up to 80% depending upon experimental conditions.[11] One important requirement is that the volume of dialysate produced does not cause breakthrough on the concentration column, prior to RP-HPLC. This method has considerable potential for rapid monitoring of residues. It appears to be particularly successful with milk and other liquid biological samples, but for tissue extracts some form of selective detection is required. Individual methods are discussed below.

Barker and co-workers have pursued a novel approach to residue analysis of both veterinary drugs and pesticides,[12] grinding samples with a C_{18}-bonded phase material, a process they term Matrix Solid Phase Dispersion (MSPD) and which is claimed uniquely to expose the cellular contents to extraction. This mixture is packed into a cartridge and eluted with solvent. MSPD has been relatively successfully applied to the analysis of sulphonamides spiked into pig muscle[13] and benzimidazoles in cattle liver.[14] The group has reported many other potential applications but if the method is to be widely accepted it will have to be extended to more typical residue concentrations and validated with incurred tissue.

[11] M. M. L. Aerts, W. M. J. Beek, and U. A. Th. Brinkman, *J. Chromatogr.*, 1990, **500**, 453.
[12] S. A. Barker, A. R. Long, and C. R. Short, *J. Chromatogr.*, 1989, **475**, 353.
[13] A. R. Long, L. C. Hsieh, M. S. Malbrough, C. R. Short, and S. A. Barker, *J. Agric. Food Chem.*, 1990, **38**, 423.
[14] A. R. Long, M. S. Malbrough, L. C. Hsieh, C. R. Short, and S. A. Barker, *J. Assoc. Offic. Anal. Chem.*, 1990, **73**, 860.

1.2   Immunochemical Methods

Immunochemical analysis of veterinary drug residues in animal tissues has been reviewed.[15-17] Methods fall into two groups, immunoassay and IAC. Immunoassay can be rapid, selective, and sensitive and has proved of considerable utility in some areas of residue analysis. In common with all analytical methods it requires considerable skill from the analyst to achieve consistent results and good LoDs. The novice user will need some time to establish the necessary technique.

Immunoassay is most efficiently applied to long sample runs of biological fluids, as for example in clinical analysis. In the hands of experienced users it is potentially capable of providing LoDs comparable to the most sophisticated instrumental methods of analysis. Immunoassays may be preferred in circumstances where analyte concentrations below $1 \mu g \, kg^{-1}$ must be measured and a mass spectrometer is not available. It is often difficult to achieve adequate selectivity at these low levels with other physico-chemical techniques.

One of the advantages frequently cited for immunoassay is speed of analysis. This arises because in many cases a relatively crude extraction procedure is adequate. It is undoubtedly possible in most cases to develop rapid immunoassays for general use where residue concentrations are $10 \mu g \, kg^{-1}$ or higher, and at lower levels where the assay is to be restricted to a single sample type. However, experience in this laboratory with hormone analysis indicates that when one assay is used to determine low levels of residue in a range of species and tissues an unacceptably high level of false positives arises unless extensive sample preparation is carried out. This was found both with an assay generated within the laboratory and when using a purchased kit. Under these circumstances the sample throughputs of immunoassay and MS methods are comparable and may actually favour the latter, particularly for multi-residue analyses where the immunoassay approach would require several different assays.

IAC clean-up for veterinary drug analysis has gained popularity only recently, although it was used for serum hormone analysis ten years ago.[18] Immobilized antibody isolates drugs cleanly from crude sample extracts prior to chromatographic or immunoassay end determination. Thus the selectivity of antibody binding may be coupled with chromatographic separation of the eluted residue enabling individual quantitation of cross-reacting metabolites and MS confirmation of analyte identity. Several different antibodies can be bound in one column to provide multi-analyte capability.

Disadvantages of current IAC methodology include a requirement for

[15] L. Fukal, *Potravin. Vedy*, 1988, **6**, 297.

[16] D. Arnold, *Lebensmittelchem., Lebensmittelqual.*, 1988, **13**, 61.

[17] R. J. Heitzman, in 'Analysis of Food Contaminants', ed. J. Gilbert, Elsevier, London, 1984, p.73.

[18] S. J. Gaskell and B. G. Brownsey, *Clin. Chem. (Winston-Salem N.C.)*, 1983, **29**, 677.

substantial amounts of antibody. Antigen–antibody complex formation is slow, and achieving complete capture of a drug requires a large excess of immobilized antibody. In addition, the gel supports currently in use are soft and subject to lengthy load–elute–wash cycle times. Silica-based matrices are under development but are not widely employed because of the problem of ensuring adequate deactivation for repeated use. One other consideration is the relatively large volume of eluent containing desorbed analyte. Some form of concentration stage is usually necessary before chromatography; if HPLC is employed this may readily be carried out by dilution with water and trapping the drug either on a reverse phase pre-column or directly on the analytical column.

Publications concerned with immunoassay will be reviewed here only where the assay may be obtained commercially or where some interesting point for residue analysis is made. More attention will be given to IAC because of the promise of the technique. Immunological methods will not become widely accepted until there is ready access to a range of products with confidence in continuity of supply. IAC columns are in principle indefinitely reusable but patterns of drug usage may change and make a product obsolete. Thus commercial production is not attractive to many potential suppliers and other sources of columns may have to be explored.

Table 4 lists commercial suppliers of immunological reagents for the analysis of veterinary drugs. This area is fast growing and the list is undoubtedly incomplete. The author would welcome information on additional products. Many drug immunoassays have been produced commercially for clinical use. Little attempt has been made to apply these to residue analysis, for reasons which may include the higher drug concentrations normally encountered in clinical practice and the expense of the highly automated instruments often required. However, gentamicin in

Table 4 *Commercial suppliers of immunochemical reagents for veterinary drug residue analysis*

---

1. Biocode Ltd.
   University Road, Heslington, York, Y01 5DE, UK
   Telephone (0904) 430616; FAX (0904) 430495
   IAC: zeranol

2. Cambridge Veterinary Sciences Ltd.
   Henry Crabb Road, Littleport, Ely, Cambridgeshire, CB6 1SE, UK
   Telephone (0353) 861911; FAX (0353) 860409
   EIA: benzylpenicillin

3. Food and Veterinary Laboratory Ltd.
   25–26 Frederick Sanger Road, The Surrey Research Park, Guildford, Surrey, GU2 5YD, UK
   Telephone (0483) 300443; FAX (0483) 301171
   ELISA: stilbenes, trenbolone and zeranol
   RIA: stilbenes, trenbolone and zeranol

Table 4 *(cont.)*

4. Genego spa.
   Via Ressel, Z.I. S. Andrea, 34170 Gorizia, Italy
   Telephone (0481) 521251; FAX·(0481) 520640
   ELISA: zeranol, stilbenes, 19-nortestosterone, trenbolone, clenbuterol
   IAC: zeranol, stilbenes, oestradiol, 19-nortestosterone and β-agonists

5. Guildhay Antisera Ltd.
   Unit 6, Riverside Business Centre, Walnut Tree Close, Guildford, Surrey, GU1
   4UG, UK
   Telephone (0483) 573727; FAX (0483) 574828
   AS: a range of drugs, including hormones and antibiotics; custom preparation

6. Idexx.
   The Old Courthouse, Hughenden Road, High Wycombe, Bucks, HP13 5DT,
   UK
   Telephone (0494) 471047; FAX (0494) 471917
   EIA: β-lactams, gentamicin, sulphamethazine, sulphadimethoxine, sulphathia-
   zole, tetracyclines, neomycin, streptomycin, phenylbutazone

7. Laboratoire d'Hormonologie
   Rue du Carmel 1, 5406 Marloie, Belgium
   Telephone (084) 31 27 02; FAX (084) 31 61 08
   AS: custom preparation
   RIA: wide range of hormones
   IAC: hormones and β-agonists

8. Novo Food Diagnostics A/S
   Frydendalsvej 30, 1809 Frederiksberg C, Copenhagen, Denmark
   Telephone 45 1 31 2440; FAX 45 1 21 7838
   ELISA: sulphamethazine

9. Randox Laboratories Ltd.
   Ardmore, Diamond Road, Crumlin, Co. Antrim, BT29 4QY, UK
   Telephone (08494) 22413; FAX (08494) 52912
   AS: many drugs; custom preparation
   ELISA: sulphamethazine
   IAC: trenbolone and 19-nortestosterone

10. Riedel-de Haën AG
    Wunstorfer Str. 40, D-3016 Seelze 1, Germany
    Telephone (0 51 37) 707-624; FAX (0 51 37) 707-123
    ELISA: chloramphenicol, clenbuterol

11. Transia
    8 Rue Saint Jean de Dieu, 69007 Lyon, France
    Telephone 72 73 03 81; FAX 72 73 43 34
    EIA: sulphamethazine, sulphadimethoxine, gentamicin, chloramphenicol, tylo-
    sin

AS   : Antisera available.
EIA   : Enzyme immunoassay – various formats.
ELISA   : Kits for enzyme-linked immunosorbent assay.
IAC   : Immunoaffinity chromatography columns.
RIA   : Kits for radioimmunoassay.
1. Hormone antisera, labels, and kits are available from many clinical assay suppliers.
2. There is rapid change in this area and the Table will undoubtedly be incomplete. The
   author would be pleased to receive details of any omissions.

kidney, muscle, and milk has been determined with the Abbott fluorescence polarization system.[19]

One major problem with all immunochemical approaches lies in obtaining the necessary antibody. Drugs are typically small molecules and do not elicit a direct antigenic response. In order to obtain antibodies it is necessary to link (conjugate) the drug covalently to a carrier protein; the animal's immune system then produces antibodies to a range of surface features of the carrier protein, including the conjugated drug. The appropriate synthetic chemistry may be difficult to carry out, particularly if the drug contains several functional groups. A 'shotgun' approach is often adopted, where conjugated products are not characterized chemically but judged only in terms of the resultant antibody. As always, failures are rarely reported. Furthermore, an animal may take from 6 months to 3 years or more to respond satisfactorily to the conjugate.

Antibody specificity is both a major advantage and disadvantage for immunochemical methods. It enables highly selective detection or isolation of analytes but at the same time complicates the development of multi-residue methods. However, specificity is lost to the region of the drug molecule involved in the protein linkage because this is sterically shielded from interaction with the immune system. If care is taken in designing conjugates, this may be used as a way of extending the antibody response to homologues or metabolites of the parent drug. In this context, unexpectedly high values obtained from immunoassays compared with chromatographic methods should be considered as a possible indication of metabolite recognition. Multi-residue analysis by immunological methods will, however, often require development of antibodies to each of the individual analytes.

1.3 Microbiological Methods

Microbiological methods are used primarily for the determination of antibiotics. Their major drawbacks lie in a reduced detection of some metabolites and in the aspecific nature of the bacterial response, which may be inhibited by numerous possible co-extractives, particularly lipids, from animal tissues. This generalized response is also a strength because it enables detection of all the members of several antibiotic families simultaneously, at low individual cost. Most microbiological methods focus on detection of penicillin and there are several families of antimicrobials which usually cannot be detected at or below MRLs, including nitrofurans and chloramphenicol.

The 'Four Plate Test' was developed for use as a screening test for the European Community[20] and is based on the inhibition of bacterial growth

---

[19] S. A. Brown, D. R. Newkirk, R. P. Hunter, G. G. Smith, and K. Sugimoto, *J. Assoc. Offic. Anal. Chem.*, 1990, **73**, 479.

[20] R. Bogaerts and F. Wolf, *Fleischwirtschaft*, 1980, **60**, 672.

in the presence of meat containing antibiotic residues. Replicate disks from a tissue sample are placed on four separate agar plates, seeded with one of two test organisms, and incubated under various conditions. False positives arise at a rate which is difficult to reduce below about 1%, partly due to interferences from long-chain fatty acids produced by microbial degradation of meat, which to some extent may be alleviated by using fresh samples.

A common cause of false positives with frozen meat, particularly pig kidney, is the presence of lysozyme, which kills the test organism. Lysozyme may be excluded by placing a semi-permeable membrane between the tissue disk and the agar plate. The Four Plate Test is often used in combination with high voltage electrophoresis to exclude false positives and help identify individual antibiotics.

There are several commercial microbial inhibition tests for antibiotics in milk. Both laboratory[21,22] and on-farm methods[23] have been reviewed recently. They are designed to be particularly sensitive to penicillin G, with LoDs of *ca.* $2 \mu g \, kg^{-1}$, ($0.003 \, IU \, ml^{-1}$), but again a selection of bacterial cultures may be used to increase the range of antibiotics detectable. It is possible to determine sulphonamides at *ca.* $15 \mu g \, kg^{-1}$, aminoglycosides including neomycin at $50–500 \mu g \, kg^{-1}$, and other antibiotics at concentrations exceeding $100 \mu g \, kg^{-1}$.

Another microbiological test format for milk analysis, the Charm II, is a whole-cell receptor assay using radiolabelled drugs to detect binding. The quantities of radioisotope used are small and are exempt from regulation in the USA. In contrast to conventional microbial inhibition assays which require several hours to complete, the receptor assay may be completed within 15 minutes. The sensitivity of the test is variable, with $2 \mu g \, kg^{-1}$ penicillin G, $1–10 \mu g \, kg^{-1}$ sulphonamide drugs, $100 \mu g \, kg^{-1}$ tetracyclines, and $50 \mu g \, kg^{-1}$ chloramphenicol claimed. The system performed acceptably under collaborative testing[24] and has been adopted as official first action by the Association of Official Analytical Chemists.

2 Applications

Few reviews of methodology exist, although an extensive survey[25] of feed additives, veterinary drugs, and residue analysis was published recently in the Japanese language. The substantial Japanese contribution to drug residue analysis is unfortunately inaccessible to most analysts and it is to be hoped that an English translation of this publication will appear. Moats has

[21] G. F. Senyk, J. H. Davidson, J. M. Brown, E. R. Hallstead, and J. W. Sherbon, *J. Food Protect.*, 1990, **53**, 158.

[22] A. H. Hands, *J. Soc. Dairy Technol.*, 1989, **42**, 92.

[23] G. M. Jones and E. H. Seymour, *J Dairy Sci.*, 1988, **71**, 1691.

[24] S. E. Charm and R. Chi, *J. Assoc. Offic. Anal. Chem.*, 1988, **71**, 304.

[25] H. Nakazawa and M. Fujitsa, *Eisei Kagaku*, 1990, **36**, 163 (*Chem Abstr.*, 1991, **114**, 4924).

assessed LC procedures for antibiotics[26,27] and Crosby[28] has summarized the problems encountered during analysis of some of the major drug categories.

In the following sections, methods for veterinary drug residue analysis are discussed, grouped according to compound type. Where reverse phase high performance liquid chromatography (RP-HPLC) is described, a C_{18} column was employed unless otherwise indicated. The term 'tissue' implies at least two of muscle, kidney, and liver, although for poultry it is also used synonymously for muscle. Other matrices are mentioned separately.

The degree of validation of method performance varies widely. A minority of workers have provided extensive documentation but in some cases recovery and precision are reported at high residue concentrations and LoDs as much as two orders of magnitude lower estimated by extrapolation. Recovery data should always be provided for concentrations in the region of the claimed LoD and the basis for its estimation made explicit.

LoDs are the result of two factors: electronic noise on the signal and interferences from the sample matrix. Where only the former is experienced the LoD is defined as three times the peak-to-peak noise. In most cases, co-extractives eluting with or near the analyte increase the LoD. The LoD is then the mean value obtained from at least 20 samples known not to contain the analyte, plus three times the standard deviation. The problems with this definition include the variation of interferences with factors such as animal breed and feed history, even when dealing with one tissue type from a single species, and also specifying what constitutes an interference in chromatographic terms.

The first problem may be dealt with by taking blanks as similar to the analytical sample as possible. Chromatographic definition of an interference relates to the resolution of the separation system employed. Co-chromatography of a supposed positive after addition of an equal amount of the analyte will eliminate factors causing variation in retention times. Consideration of the peak profiles of real chromatograms indicates that an interference of similar concentration to the analyte may be difficult to detect visually if its retention time differs from that of the analyte by less than three quarters of the peak width at half-height.

Thus the area of the chromatogram over which noise should be measured for the purpose of determining LoDs is the anticipated retention time plus or minus one peak width at half height. In terms of ideal gaussian peak shapes this is equivalent to about four standard deviations. When working at or near the LoD, baseline noise can cause problems when determining peak areas by integration. In this case, peak height measurement may prove more accurate.

[26] W. A. Moats, in 'Agricultural Uses of Antibiotics', ACS Symposium Series, No. 320, American Chemical Society, Washington, DC, 1986, p. 154.
[27] W. A. Moats, *J. Assoc. Offic. Anal. Chem.*, 1990, **73**, 343.
[28] N. T. Crosby, *J. Assoc. Public Anal.*, 1986, **24**, 111.

The terms 'limit of determination' or 'limit of decision' are sometimes used. The former should reflect the lowest level at which the analysis has been validated; the latter indicates a level below which any result is not of interest, possibly for example the MRL. Again unfortunately they are often not defined.

2.1 Multi-residue Methods

Most chromatographic methods are specific to individual residues; the problems of achieving residue-level LoDs have been outlined above and they are correspondingly greater if the simultaneous determination of several structurally unrelated compounds is required. However, certain drug classes are relatively easy to co-extract and purify. Sulphonamides, nitrofurans, chloramphenicol, and trimethoprim (TMP) are readily determined in a single assay but several important groups of compounds, including hormones, β-lactams, β-agonists, and ionophores, cannot effectively be incorporated into these schemes. Except for sulphonamides, detection limits are often higher in multi-residue methods than for dedicated analyses.

Petz described the analysis in one hour of five sulphonamides together with chloramphenicol and furazolidone (FZD) in eggs, meat, and milk,[29] although detection limits for the latter two drugs were not adequate to meet current MRLs, a problem common to most multi-residue methods. Nose and colleagues devised[30] a method for four sulphonamides, four nitrofurans, and three quinolone carboxylic acids in cultured fish, including eel and rainbow trout, with LoDs for all analytes of 20–80 μg kg^{-1}. An acetone extraction was employed and recoveries were 71–92% at 500 μg kg^{-1}. The same workers used acidic extraction for monitoring a total of 11 drugs including sulphonamides, tetracyclines, chloramphenicol, and nitrofurans with similar recoveries and LoDs.[31]

Nagata and Saeki[32] determined 17 antimicrobials (clopidol, TMP, ormetoprim, pyrimethamine, FZD, dinitolmide, difurazone, nicarbazin, ethopabate, decoquinate, and seven sulphonamides) in chicken tissue. Residues were extracted with warm methanol and the evaporated extract dissolved in pH 5.5 buffer, defatted with hexane, and partitioned into dichloromethane. RP-HPLC was carried out after alumina clean-up. Average recoveries were greater than 65% at 200 μg kg^{-1} and LoDs 30–50 μg kg^{-1}. Hori[33] used a very similar method for 11 of these compounds in chicken tissue and claimed the same LoDs.

[29] M. Petz, *Z. Lebensm.-Unters. Forsch.*, 1983, **176**, 289.

[30] N. Nose, Y. Hoshino, Y. Kikuchi, M. Horie, K. Saitoh, T. Kawachi, and H. Nakazawa, *J. Assoc. Offic. Anal. Chem.*, 1987, **70**, 714.

[31] M. Horie, Y. Hoshino, N. Nose, H. Iwasaka, and H. Nakazawa, *Eisei Kagaku*, 1985, **31**, 371.

[32] T. Nagata and M. Saeki, *Shokuhin Eiseigaku Zasshi*, 1988, **29**, 13 (*Chem. Abstr.*, 1988, **109**, 36 722).

[33] Y. Hori, *Shokuhin Eiseigaku Zasshi*, 1983, **24**, 447 (*Chem. Abstr.*, 1984, **100**, 66 681).

Barker and colleagues have demonstrated that multiple residues may be determined using the MSPD approach.[12] Direct HPLC analysis was possible of benzimidazole and penicillins fractionally eluted from prepared MSPD cartridges containing cattle muscle spiked at 200 μ kg^{-1}.

Malisch[34] has described the determination of sulphonamides, nitrofurans, chloramphenicol, and TMP in eggs, milk, and tissue by acetonitrile extraction and extensive liquid–liquid partitioning. Malisch extended this work to an ambitious RP-HPLC multi-residue scheme.[35-38] Up to 27 sulphonamides (LoD typically 20 μg kg^{-1}) could be determined simultaneously.[35] Final extracts are split and analysed by a number of RP-HPLC gradient elution separations with UV wavelength programming. This approach is time-consuming but appears to be successful and should be considered if a number of analyses are required from a sample. With the exception of chloramphenicol (which if required is determined separately by GC–ECD), LoDs are in the region of MRLs although additional sensitivity would be desirable for nitro compounds. The problems of using diode array techniques for partial peak identity confirmation were discussed.[38]

2.2 Aminoglycosides

Shaikh and Allen[39] have presented a thorough literature survey of non-residue chemical methods for aminoglycosides, which includes much useful information on the properties and chromatography of these drugs. Aminoglycosides (see Figure 1) are highly polar compounds with poor native chromophore characteristics; one additional problem is that some of the drugs are mixtures of several closely related compounds. Analysis in tissue is difficult and few methods are available. Further work is required to improve LoDs.

Aminoglycosides are heat, acid, and base stable, but should be handled in plastic labware as they may be adsorbed on to glass surfaces.[40] Gentamicin standard solution was not significantly affected[19] by heating for up to 4 hours at 70 °C within the pH range 3.4–11.5. Kidney is the target organ[40] and extraction must be carried out in basic buffer.[40,41] HPLC detection is usually by fluorimetry of o-phthalaldehyde (OPA) derivatives.

Agarwal[41] determined gentamicin at 200 μg kg^{-1} in cattle muscle by extraction into basic buffer, which was deproteinized by boiling. Clean-up was performed by ion exchange chromatography and the drug isolated on a

[34] R. Malisch, U. Sandmeyer, and K. Kypko-Hutter, *Lebensmittelchem. Gerichtl. Chem.*, 1984, **38**, 11.
[35] R. Malisch, *Z. Lebensm.-Unters. Forsch.*, 1986, **182**, 385.
[36] R. Malisch, *Z. Lebensm.-Unters. Forsch.*, 1986, **183**, 253.
[37] R. Malisch, *Z. Lebensm.-Unters. Forsch.*, 1987, **184**, 467.
[38] R. Malisch and L. Huber, *J. Liq. Chromatogr.*, 1988, **11**, 2801.
[39] B. Shaikh and E. H. Allen, *J. Assoc. Offic. Anal. Chem.*, 1985, **68**, 1007.
[40] B. Shaikh, E. H. Allen, and J. C. Gridley, *J. Assoc. Offic. Anal. Chem.*, 1985, **68**, 29.
[41] V. K. Agarwal, *J. Liq. Chromatogr.*, 1989, **12**, 613.

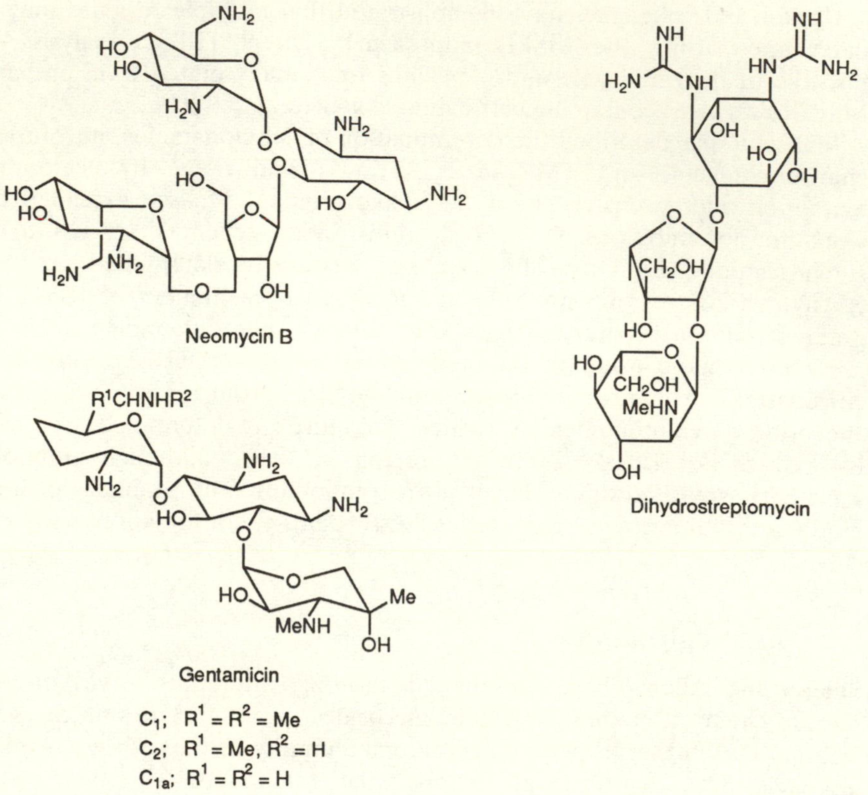

Neomycin B

Dihydrostreptomycin

Gentamicin

C_1: $R^1 = R^2 = Me$
C_2: $R^1 = Me, R^2 = H$
C_{1a}: $R^1 = R^2 = H$

Figure 1

silica solid-phase extraction (SPE) cartridge and derivatized *in situ* with OPA before ion-pair HPLC. A development[42] of this method involved isolation of neomycin from milk, without defatting or deproteinizing, on to an ion exchange resin and *in situ* derivatization with OPA. A protein-resistant RP-HPLC column was employed. Recovery was 100% at 100 $\mu g\,kg^{-1}$ and the LoD was 50 $\mu g\,kg^{-1}$.

Shaikh and his co-workers measured neomycin in kidney and muscle from cattle and pig.[40] Extracts were deproteinized by boiling and centrifuged but no additional clean-up was employed; acidified supernatant was analysed directly by ion-pair HPLC with post-column addition of OPA. The LoD was in the region of 1000 $\mu g\,kg^{-1}$. A similar procedure was applied to milk.[43] Defatted, deproteinized milk was injected directly into the HPLC system, permitting an LoD of about 150 $\mu g\,kg^{-1}$.

Farrington and colleagues[44] monitored lincomycin residues at 50 $\mu g\,kg^{-1}$

[42] V. K. Agarwal, *J. Liq. Chromatogr.*, 1990, **13**, 2475.
[43] B. Shaikh and J. Jackson, *J. Liq. Chromatogr.*, 1989, **12**, 1497.
[44] W. H. H. Farrington, S. D. Cass, A. L. Patey, and G. Shearer, *Food Add. Contam.*, 1988, **5**, 67.

in cattle and pig kidney, using a clean-up including HPLC prior to silylation with bis(trimethylsilyl)trifluoroacetamide (BSTFA) and GC–NPD determination. Recoveries were rather variable. Streptomycin was determined[45] at a claimed LoD of 500 μg kg^{-1} in meat by an ion-pair extraction and HPLC with post-column ninhydrin reaction detection. Destomycin A and hygromycin B were analysed[46] in pig muscle at 100 μg kg^{-1} and 300 units kg^{-1} respectively, using trichloroacetic acid (TCA) extraction, ion exchange clean-up, and HPLC post-column reaction with OPA.

The Abbott fluorescence polarization immunoassay system has been employed[19] to determine gentamicin in milk and cattle and sheep kidney and muscle at levels of 10–20 μg kg^{-1}. Initial tissue extraction was found to be performed best by digestion of unhomogenized material with sodium hydroxide. Recovery was significantly worse from skeletal muscle (70%) than kidney (90%) and it was suggested that additional homogenization or protease treatment should be employed.

2.3 Benzimidazoles

Few methods for benzimidazoles (Figure 2) are available, despite the large-scale use of these compounds as anthelmintics. Metabolism of the thio-substituted compounds consists variously of sulphide oxidation first to sulphoxide and then to sulphone, aryl hydroxylation, or *N*-deacetylation.[47,48] The target organ is liver.

A method[49] for the determination of fenbendazole (FBZ), oxfendazole (OFZ), thiabendazole (TBZ), and 5-hydroxythiabendazole (HTBZ) in milk at 5 μg kg^{-1} has been described. Partitioning of benzimidazoles between organic and aqueous phases is controlled by pH adjustment, permitting extraction and defatting, followed by SPE clean-up on silica. BHT was added during the extraction to prevent additional oxidation of the residues; this was particularly important for recovery of HTBZ. Separation by RP-HPLC using different combinations of columns and methanol–pH 7.0 ammonium phosphate buffers was followed with UV detection at wavelengths of 298 nm, or, for HTBZ, 318 nm.

A similar method, using solid-phase extraction, was devised for determination of eight benzimidazoles in cattle liver.[7] Flow-programmed RP-HPLC at 290 nm gave good to excellent recoveries at levels close to the LoDs of 10–30 μg kg^{-1}. Non-homogeneity of residue distribution in the liver was considered probable. The same group examined conditions[50] for

[45] A. Okayama, Y. Kitada, Y. Aoki, S. Umesako, H. Ono, Y. Nishii, and H. Kubo, *Bunseki Kagaku*, 1988, **37**, 221 (*Chem. Abstr.*, 1988, **109**, 36 714).

[46] K. Nakaya, A. Sugitani, and M. Kawai, *Shokuhin Eiseigaku Zasshi*, 1987, **28**, 487 (*Chem. Abstr.*, 1988, **109**, 5359).

[47] P. Delatour and R. Parish, in 'Drug Residues in Animals', ed. A. G. Rico, Academic Press, New York, 1986, p. 65.

[48] D. H. Watson, *Food Chem.*, 1983, **12**, 167.

[49] S. S.-C. Tai, N. Cargile, and C. J. Barnes, *J. Assoc. Offic. Anal. Chem.*, 1990, **73**, 368.

[50] A. R. Long, L. C. Hsieh, C. R. Short, and S. A. Barker, *J. Chromatogr.*, 1989, **475**, 404.

HPLC of benzimidazoles and presented UV spectra of seven compounds. They also determined FBZ, together with OFZ and other metabolites, in goat faeces by direct probe EI-MS and EI-MS–MS at LoDs of 200–450 μg kg^{-1}. GC separation, even of the derivatized compounds, could not be achieved.[51] TBZ and HTBZ were analysed[52] by GC–MS using on-column methylation at 100 μg kg^{-1} with good recovery from cattle and pig tissues, including cattle fat.

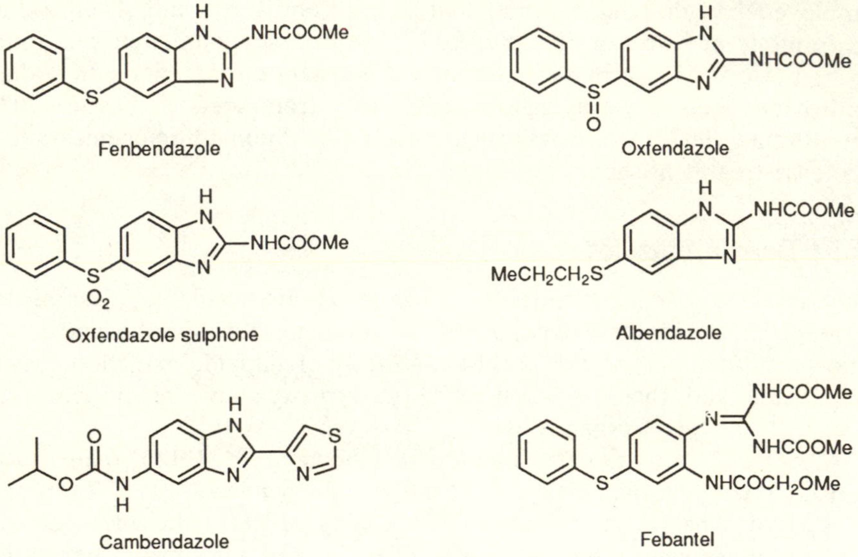

Figure 2

Farrington *et al.*[53] reported the determination of six benzimidazoles in sheep or chicken liver by extraction into ethyl acetate followed by sequential SPE clean up; first loading an acidic alumina cartridge in hexane:chloroform (25:75) with elution by methanol, then a C$_{18}$ cartridge, loaded in methanol:water (25:75), eluted by acetonitrile. Separation was by RP-HPLC and detection used UV (290 nm; OFZ and mebendazole) or fluorescence (312/355 nm; TBZ, cambendazole and albendazole sulphone). Recoveries at 20 μg kg^{-1} (the LoD) and 50 μg kg^{-1} were 70–90%.

The MSPD technique has been shown to recover five and seven

[51] S. A. Barker, L. C. Hsieh, T. R. McDowell, and C. R. Short, *Biomed. Environ. Mass Spectrom.*, 1987, **14**, 161.
[52] W. J. A. VandenHeuvel, J. S. Wood, M. DiGiovanni, and R. W. Walker, *J. Agric. Food Chem.*, 1977, **25**, 386.
[53] W. H. H. Farrington, S. Chapman, and D. Tyler, in Proc. Euroresidue Conf. Residue Vet. Drugs Food, ed. N. Haagsma, A. Ruiter, and P. B. Czedik-Eysenberg, University of Utrecht, Utrecht, 1990, p. 185.

benzimidazoles from cattle liver[14] and milk[54] respectively. A simple additional clean-up on alumina was required for the liver samples. RP-HPLC analysis used 17 mM phosphoric acid–acetonitrile (6 + 4) and UV detection at 290 nm. LoDs of less than 100 μg kg^{-1} were estimated. Some oxidation of fenbendazole during the procedure is apparent from the recovery figures given.

Triclabendazole and its oxidation products were determined[55] in goat milk by deproteinization with acetone and C$_{18}$ SPE. Analysis by RP-HPLC with a pH 6.7 buffered mobile phase and fluorescence detection at 300/676 nm provided LoDs of 20–40 μg kg^{-1}.

2.4 β-Agonists

Although licensed only for respiratory disorders in cattle and horses, β-agonists (Figure 3) have been used illegally to increase the lean/fat ratio in livestock. The β-blocker carazolol will be considered together with the tranquilizers because of the multi-analyte methods available. Residues of clenbuterol in cattle liver have been implicated as the cause of a number of cases of ill-health in Spain.[56] These drugs fall into two major classes: substituted anilines, including clenbuterol, and substituted phenols, such as salbutamol. This distinction is important because most methods for the former category depend upon pH manipulation to partition the drugs alternately between organic and aqueous phases. They are therefore not valid for the phenolic compounds which are charged under all pH conditions. β-Agonists are known to be protein bound and tissue pretreatment with subtilisin or other protease is obligatory. Liver is the target organ for clenbuterol but the drug is cleared rapidly, with a half-life of about 36 h, and LoDs of 1 μg kg^{-1} or lower are required.

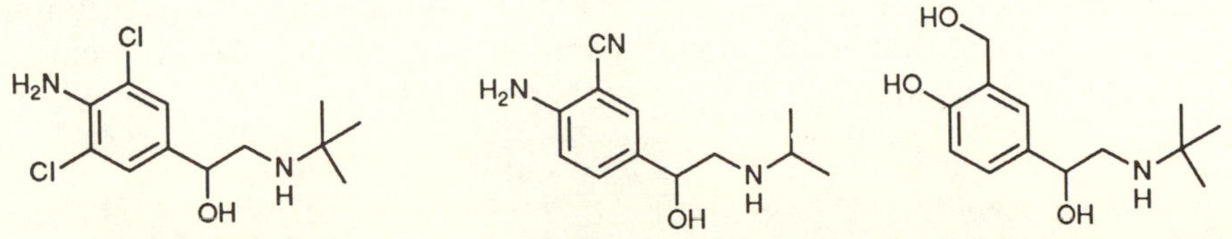

Clenbuterol

Cimaterol

Salbutamol
(Albuterol)

Figure 3

[54] A. R. Long, L. C. Hsieh, M. S. Malbrough, C. R. Short, and S. A. Barker, *J. Assoc. Offic. Anal. Chem.*, 1989, **72**, 739.
[55] L. D. B. Kinabo and J. A. Bogan, *J. Vet. Pharmacol. Ther.*, 1988, **11**, 254.
[56] J. F. Martinez-Navarro, *Lancet*, 1990, **336**, 1311.

Clenbuterol has pK_as of <1 and 9.6. Forster *et al.*[57] described a method for clenbuterol in plasma, which was stated to be applicable to proteolysed tissue. Plasma is mixed with pH 9.5 buffer and partitioned against t-butylmethyl ether (TBME). Clenbuterol is back extracted into 0.2 M sulphuric acid which is washed with hexane and TBME and then made alkaline before final re-extraction into TBME. The final determination was by capillary GC–ammonia chemical ionization-mass spectrometry (GC–CI–MS). CI was chosen because the electron impact (EI) spectrum of clenbuterol is dominated by the base peak at m/z 86, with suitable higher mass fragments giving rise to no more than a few percent of base peak intensity. However, this ion arises from the t-butyl end of the molecule and is surprisingly free from interference, permitting single ion monitoring at levels of about 1 μg kg^{-1}. A similar method was developed[58] for low-level (0.01 μg kg^{-1}) residues. Deuterated internal standards are added to provide compensation for losses during clean-up and other manipulations.

An alternative method[59] involves HPLC separation of the final residue with post-column derivatization of clenbuterol, which requires the addition of three reagents. The method appears to work well, with an LoD of below 1 μg kg^{-1}, but the confirmation available from MS detection will ensure that this is the method of choice. Clenbuterol possesses UV maxima at about 240 and 300 nm in aqueous media, with the former having three times the extinction coefficient of the latter. A simplified clean-up, involving partition from alkali-treated tissue into hexane and back extraction into acid followed by detection at 240 nm, enables determination of clenbuterol at 2–3 μg kg^{-1}; a co-extractive from liver may restrict the LoD.[60]

An IAC/GC–MS[61] method provided 1 μg kg^{-1} LoDs in liver of clenbuterol, cimaterol, terbutaline, and salbutamol. Sample preparation prior to IAC consisted of extraction into buffer, adjustment to pH 10 and C_{18} SPE. For urine samples, on-line IAC–HPLC analysis has been reported.[62] The LoD was given as 0.5 μg kg^{-1}, with a recovery of 82% for urine spiked at 5 μg kg^{-1}. This method has obvious potential for extension to tissue analysis. It is interesting that these workers and others have found clenbuterol not to be released from the IAC column by aqueous alcohols, unlike hormone–antibody immunoaffinity complexes. Acetic acid at 0.01–0.1 M is required.

[57] H. J. Forster, K. L. Rominger, E. Ecker, H. Peil, and A. Wittrock, *Biomed. Environ. Mass Spectrom.*, 1988, **17**, 417.
[58] J. Girault and J. B. Fourtillan, *J. Chromatogr.*, 1990, **518**, 41.
[59] J.-M. Degroodt, B. Wyhowski de Bukanski, H. Beernaert, and D. Courtheyn, *Z. Lebensm. Unters. Forsch.*, 1989, **189**, 128.
[60] R. Parker and I. C. Shaw, personal communication.
[61] R. Schilt, W. Haasnoot, M. A. Jonker, H. Hooijerink, and R. J. A. Paulussen, in Proc. Euroresidue Conf. Residue Vet. Drugs Food, ed. N. Haagsma, A. Ruiter, and P. B. Czedik-Eysenberg, University of Utrecht, Utrecht, 1990, p. 320.
[62] W. Haasnoot, M. E. Ploum, R. J. A. Paulussen, R. Schilt, and F. A. Huf *J. Chromatogr.*, 1990, **519**, 323.

Few methods exist for other β-agonists in tissue. Salbutamol was measured in organs from dogs[63] and terbutaline in human tissues.[64] These procedures might perhaps be extended to animal foodstuffs. Salbutamol has a much lower UV extinction coefficient than does clenbuterol but is readily detected by fluorescence with excitation and emission at 220 and 305 nm.

2.5 β-Lactams

Current methods for the chemical analysis of β-lactam residues are generally insensitive and slow; most methods focus on penicillins in milk. Very little work has been carried out on cephalosporins. Selective structures are shown in Figure 4. Moats[26,27] has reviewed tissue analysis. Dilute aqueous solutions of β-lactams degrade relatively rapidly. Ampicillin solutions held at 10 °C were most stable at pH 5–6. At pH 3 losses of over 50% were reported after 18 days.[65] A reduction of 25% in procaine penicillin residues in chicken muscle was noted[66] on storage at −18 °C for 3 months. β-Lactams bind to proteins but are readily released with denaturing organic solvents.[67]

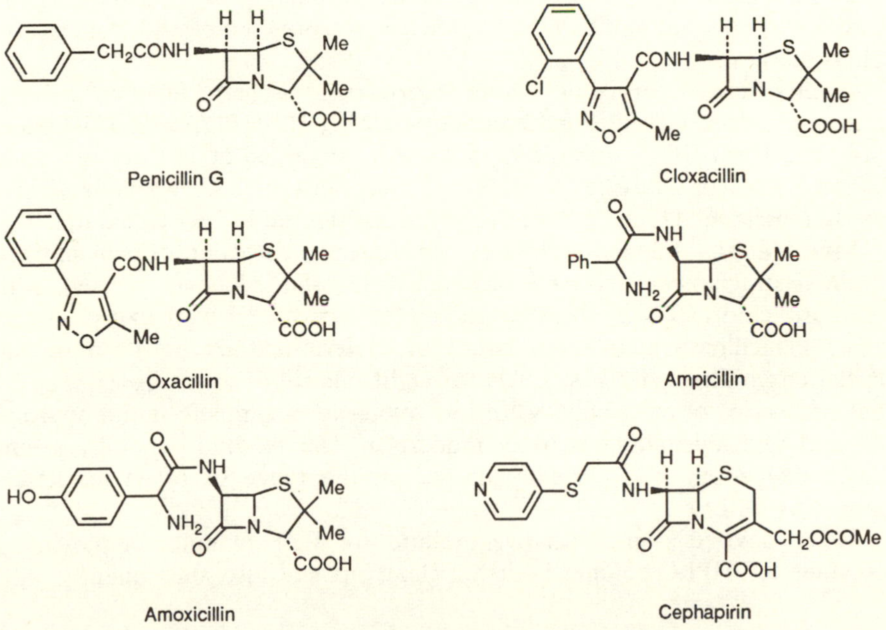

Penicillin G

Cloxacillin

Oxacillin

Ampicillin

Amoxicillin

Cephapirin

Figure 4

[63] M. C. Saux, J. Girault, S. Bouquet, J. B. Fourtillan, and Ph. Courtois, *J. Pharmacol.*, 1986, **17**, 692.

[64] J. G. Leferink, I. Wagemaker-Engels, and R. A. A. Maes, *J. Anal. Toxicol.*, 1978, **2**, 86.

[65] T. Nagata and M. Saeki, *J. Assoc. Offic. Anal. Chem.*, 1986, **69**, 448.

[66] A. Calles, B. Moreno, A. Otera, and M. L. Garcia, *An. Bromatol.*, 1988, **40**, 133 (*Chem. Abstr.* 1989, **110**, 74005).

[67] K. Tyczkowska, R. D. Voyksner, and A. L. Aronson, *J. Chromatogr.*, 1989, **490**, 101.

Some measure of confirmation of penicillins may be obtained by analysis of samples before and after treatment with β-lactamase. Typical LoDs for benzylpenicillin (BZP) in milk or tissues are 5–50 μg kg^{-1} (1 μg = 1.67 IU) and analytical throughput is no more than 5–10 samples per day per analyst. Also, the conditions for clean-up limit many of these methods to monobasic penicillins, excluding drugs such as amoxicillin and ampicillin.[68,69]

Most methods for β-lactams are based on RP-HPLC, but determination of penicillins is handicapped by the absorption properties of BZP, which has UV maxima in the 250–265 nm range with low extinction coefficients. Berger and Petz[70] have demonstrated the potential for determination of seven penicillins in milk at LoDs of about 1 μg kg^{-1} by derivatization with the fluorescence label 4-bromomethy1-7-methoxycoumarin.

Moats has developed methods for BZP and cloxacillin in milk[68] (2–5 μg kg^{-1}) and cattle and pig tissue[69] (50–500 μg kg^{-1}) with initial extraction into pH 2.2 buffer to suppress the carboxylate ionization. This was followed by liquid–liquid partition, reverse phase chromatography, and detection by UV at 220 nm. Much higher concentrations of penicillins were found in incurred tissue with the HPLC procedure compared with microbiological methods. It was suggested that tissue binding reduces the availability of the antibiotic in the tissue cores employed for microbiological testing.

A procedure[71] for cattle tissue incorporated a basic alumina column clean-up before C_{18} SPE and column-switching RP-HPLC with UV detection at 210 nm. Recoveries of 75–90% were obtained at 1000 μg kg^{-1}; the claimed LoD of 50 μg kg^{-1} seems rather optimistic. BZP was spiked into cattle muscle at 200 μg kg^{-1} and recovered using the MSPD technique.[12]

Munns *et al.*[72] adopted a different approach to detection of penicillins in milk. Extracts were cleaved enzymatically, the products treated with mercuric chloride, and liberated aldehydes derivatized with dansyl hydrazine. Penicilloate metabolites could be determined by omission of the initial enzymatic hydrolysis. LoDs for eight penicillins were 20–50 μg kg^{-1}, but recoveries of cloxacillins were low owing to incomplete initial hydrolysis and reproducibility was only moderate. The method generally seems cumbersome but it is currently unique in the range of β-lactams determined by HPLC.

Moats devised[73] an alternative method for BZP in milk, employing a polymer RP-HPLC column both for clean up and also subsequently with

[68] W. A. Moats, *J. Agric. Food Chem.*, 1983, **31**, 880.

[69] W. A. Moats, *J. Chromatogr.*, 1984, **317**, 311.

[70] K. Berger and M. Petz, in 'Proc. Euroresidue Conf. Residue Vet. Drugs Food', ed. N. Haagsma, A. Ruiter, and P. B. Czedik-Eysenberg, University of Utrecht, Utrecht, 1990, p. 118.

[71] H. Terada, M. Asanoma, and Y. Sakabe, *J. Chromatogr.*, 1985, **318**, 299.

[72] R. K. Munns, W. Shimoda, J. E. Roybal, and C. Vieira, *J. Assoc. Offic. Anal. Chem.*, 1985, **68**, 968.

[73] W. A. Moats, *J. Chromatogr.*, 1990, **507**, 177.

an acidic mobile phase for analysis. An LoD in the region of $10 \mu g \, kg^{-1}$ was achieved. Although part of the clean-up is automated the method is relatively slow, requiring 4–5 hours per sample. The clean up injection volume was 2 ml, which took 27 minutes to load with the autosampler used, and it was stated that a larger volume would have been preferred. However, at least one instrument (Gilson) will handle unlimited volumes in a much shorter time. Ampicillin in fish[65] was extracted with methanol, cleaned up on a deactivated Florisil SPE cartridge, and determined by RP-HPLC with methanol–pH 6.0 buffer and UV detection at 222 nm, affording an LoD of about $50 \mu g \, kg^{-1}$.

Two relatively simple methods for milk have been reported. BZP, ampicillin and phenoxypenicillin were extracted by passage of undiluted milk through a C_{18} SPE cartridge which was washed with crown ether in saline methanol and eluted with methanol.[74] Residues were analysed by ion-pair HPLC, with claimed LoDs of $30 \mu g \, kg^{-1}$. Cephapirin in milk[75] at $20 \mu g \, kg^{-1}$ was determined by a similar procedure. The cephalosporin was detected by UV at 254 nm after RP-HPLC. The column required purging with strong solvent after each injection and ion-pair separation was found to reduce sensitivity. Another method[67] for BZP in milk consisted of dilution with solvent, ultrafiltration, and direct ion-pair RP-HPLC. An LoD of $10 \mu g \, kg^{-1}$ was reported. Thermospray LC–MS was capable of confirming levels down to $100 \mu g \, kg^{-1}$.

A recent GC–NPD procedure[76] for seven penicillins in milk (and tissues when an extended clean-up was employed) at sub-$\mu g \, kg^{-1}$ concentrations was described by Meetschen and Petz. The penicillin residues are derivatized with diazomethane. This seems the best of the methods available for analysis of milk. It is, however, still slow and requires specialized GC equipment. These workers highlight some of the problems encountered when working with penicillins: acid instability of some drugs and potential considerable losses when evaporating solutions containing sub-μg amounts of BZP.

In general microbiological methods are employed for screening foods for β-lactam residues, particularly in the dairy industry. Although the sensitivity of microbiological methods is adequate, there are still problems in response time and there is no indication of the individual β-lactam implicated in a positive result. Analysis time is shorter for the radio-receptor assay[24] but again only group identification is possible. Several manufacturers have introduced antibody-based methods in an attempt to address these constraints (see Table 4). Jackman *et al.*[77] have reported that one commercial assay may be completed in 15 minutes and detects $6 \mu g \, kg^{-1}$ BZP in milk with 99% confidence and $3 \mu g \, kg^{-1}$ with 75–95% confidence.

[74] H. Terada and Y. Sakabe, *J. Chromatogr.*, 1985, **348**, 379.
[75] A. I. MacIntosh, *J. Assoc. Offic. Anal. Chem.*, 1990, **73**, 880.
[76] U. Meetschen and M. Petz, *J. Assoc. Offic. Anal. Chem.*, 1990, **73**, 373.
[77] R. Jackman, J. Chesham, S. J. Mitchell, and S. D. Dyer, *J. Soc. Dairy Technol.*, 1990, **43**, 93.

No other penicillin cross-reacted at more than 6%, and results on 1651 milk samples correlated relatively well with the microbiological Delvotest.

2.6   Carbadox and Olaquindox

Both compounds are rapidly metabolized to monoxy and desoxy compounds (see Figure 5), residues occurring in kidney and liver;[78] little change was seen[78] in levels during storage at −20 °C for 4 weeks. The final product from carbadox[79] and the one most often determined is quinoxaline-2-carboxylic acid (QCA). The target organ for QCA is liver; muscle residue was not found.[80] Carbadox and olaquindox are light-sensitive compounds and sample manipulations should be performed only under the minimum of indirect incandescent illumination.[81] All glassware should be ambered. Carbadox is only very sparingly soluble in common solvents. The preparation of concentrated standards in DMF and the UV spectra of the parent compounds and metabolites have been reported.[81,82]

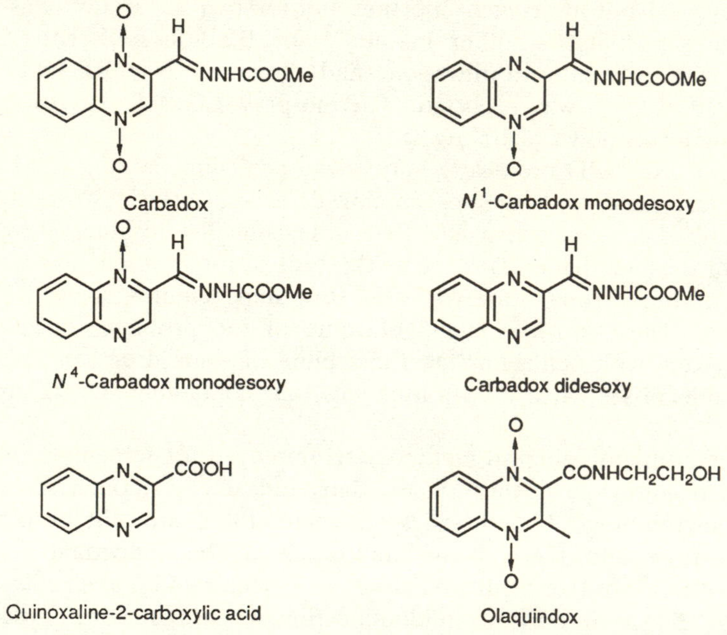

Carbadox

N^1-Carbadox monodesoxy

N^4-Carbadox monodesoxy

Carbadox didesoxy

Quinoxaline-2-carboxylic acid

Olaquindox

Figure 5

[78] A. I. MacIntosh, G. Lauriault, and G. A. Neville, *J. Assoc. Offic. Anal. Chem.*, 1985, **68**, 665.
[79] R. Ferrando, R. Truhaut, J. P. Raynaud, and J. P. Spanoghe, *Toxicology*, 1975, **3**, 369.
[80] M. G. Lauridsen, C. Lund, and M. Jacobsen, *J. Assoc. Offic. Anal. Chem.*, 1988, **71**, 921.
[81] A. I. MacIntosh and G. A. Neville, *J. Assoc. Offic. Anal. Chem.*, 1984, **67**, 958.
[82] M. M. L. Aerts, W. M. J. Beek, H. J. Keukens, and U. A. Th. Brinkman, *J. Chromatogr.*, 1988, **456**, 105.

Both drugs are amenable to extraction in polar solvents, clean-up with alumina, and RP-HPLC using UV detection at 350 nm. Carbadox and nitrofurazones were analysed at 10–40 μg kg^{-1} in pig tissue.[81] Olaquindox was determined in pig muscle and liver[83] with an estimated LoD of 20 μg kg^{-1}. Recoveries at 50 μg kg^{-1} were 60%. A method for both drugs at 10 μg kg^{-1}, developed in this laboratory,[84] consisted of ethyl acetate extraction, defatting, clean-up on C$_{18}$ SPE, and gradient RP-HPLC.

QCA has been analysed[80] using packed column GC at 10 μg kg^{-1} in pig liver and muscle. LoDs should be improved by using capillaries. MS data[85] have been reported. Aerts and co-workers[82] developed a rapid method for carbadox and its three reduced metabolites with a simple clean-up, and employed post-column reaction with sodium hydroxide to obtain sensitive and specific detection. Evaporation of extracts to dryness caused losses, but LoDs of *ca.* 2 μg kg^{-1} were obtained. A fully automated system incorporating dialysis was reported[82] to give LoDs of 25 μg kg^{-1}. Improved limits could probably be obtained with current dialysis instruments.

2.7   Chloramphenicol

The toxicology of chloramphenicol (CAP) and methods for residue analysis were comprehensively reviewed by Allen[86] in 1985. The drug (Figure 6) is metabolized to the 3-glucuronide, but MRLs are stated in terms of the parent drug and enzymolysis is therefore not routinely carried out. Some binding to egg proteins[87] has been noted but other workers have not experienced problems. Standard solutions in water are of only moderate stability. Few methods are available for the analogue thiamphenicol; it has been determined[88] with an estimated LoD of 100 μg kg^{-1} in bovine plasma by solvent extraction and RP-HPLC with UV detection at 224 nm, at which wavelength the chromatograms were surprisingly clean.

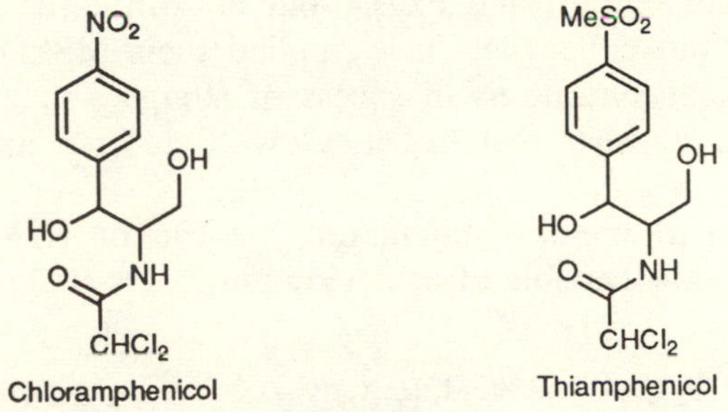

Chloramphenicol   Thiamphenicol

Figure 6

[83] T. Nagata and M. Saeki, *J. Assoc. Offic. Anal. Chem.*, 1987, **70**, 706.
[84] W. H. H. Farrington, A. L. Patey, J. Bygrave, J. M. Croucher, and G. Shearer, unpublished results.
[85] M. J. Lynch and S. R. Bartolucci, *J. Assoc. Offic. Anal. Chem.* 1982, **65**, 66.
[86] E. H. Allen, *J. Assoc. Offic. Anal. Chem.*, 1985, **68**, 990.
[87] L. Weber, *J. Chromatogr. Sci.*, 1990, **28**, 501.
[88] L. Felice, E. H. Abdennebi, and M. Ashraf, *J. Assoc. Offic. Anal. Chem.*, 1988, **71**, 1156.

Virtually every chromatographic and immunochemical technique has been employed at some time for CAP analysis. Allen[86] described 24 chromatographic methods, equally split between HPLC–UV (275 nm), GC–ECD, and TLC, where detection was typically by reduction of CAP with stannous chloride and reaction with 4-dimethylaminobenzaldehyde. Thiamphenicol, meta-CAP, and monochloro-CAP have been used as internal standards. RIA, EIA, and IAC methods are also available. CAP has been included in several multi-analyte schemes, but not with an adequate LoD (see Section 2.1).

A rapid HPLC method for CAP in muscle[89] developed by Keukens *et al.* was successfully collaboratively tested.[90] It was stated to have been in routine use over three years and several hundred samples. Throughput was 20 per analyst per day, recovery was 55% at $10 \, \mu g \, kg^{-1}$, and the LoD was $1.5 \, \mu g \, kg^{-1}$. Parker and Shaw[91] encountered losses of CAP during analysis of liver by this method, which they ascribed to continued oxidation of the drug by co-extracted cytochrome P_{450} during sample clean-up. Incorporation of piperonyl butoxide, a P_{450} inhibitor, into the water used for extraction was found to double recovery at $20 \, \mu g \, kg^{-1}$ to 61%. The use of water may render this method particularly susceptible to continued enzymatic oxidation of CAP. Other approaches frequently employ extraction by organic solvents where there seems little problem in obtaining reasonably high recoveries.[86] Samples analysed during the collaborative trial were found[90] to contain CAP concentrations near those expected, so residues are undoubtedly stable, at least in the medium term.

Determination of CAP at levels down to $0.025 \, \mu g \, kg^{-1}$ in tissue, milk, and eggs was carried out by negative ion MS[92] following the Malisch multi-residue procedure – extraction with acetonitrile, defatting with hexane, addition of sodium chloride, and partition into dichloromethane, followed by silica gel column clean-up and silylation. CAP has been determined simultaneously with sulphonamides and FZD,[29] and with tetracyclines, sulphonamides, and FZD[31] but in both cases with relatively high LoDs. Long and colleagues have applied their MSPD technique to CAP in milk[93] at concentrations in excess of $60 \, \mu g \, kg^{-1}$; examination of the chromatograms suggests that LoDs below $10 \, \mu g \, kg^{-1}$ may be difficult to achieve.

Arnold and co-workers have compared the use of RIA and GC[94,95] methods. The RIA was capable of achieving low LoDs ($0.2 \, \mu g \, kg^{-1}$) and a

[89] H. J. Keukens, W. M. J. Beek, and M. M. L. Aerts, *J. Chromatogr.*, 1986, **352**, 445.

[90] R. M. L. Aerts, H. J. Keukens, and G. A. Werdmuller, *J. Assoc. Offic. Anal. Chem.*, 1989, **72**, 570.

[91] R. M. Parker and I. C. Shaw, *Analyst (London)*, 1988, **113**, 1875.

[92] P. Fürst, C. Krüger, H. A. Meemken, and W. Groebel, *Dtsch. Lebensm.-Rundsch.*, 1988, **84**, 103.

[93] A. R. Long, L. C. Hsieh, A. C. Bello, M. S. Malbrough, C. R. Short, and S. A. Barker, *J. Agric. Food Chem.*, 1990, **38**, 427.

[94] D. Arnold and A. Somogyi, *J. Assoc. Offic. Anal. Chem.*, 1985, **68**, 984.

[95] G. Balizs and D. Arnold, *Chromatographia*, 1989, **27**, 489.

wealth of data was presented on its performance characteristics. Other immunochemical methods have been developed by van de Water and Haagsma. An ELISA format immunoassay[96] was produced for analysis of pig muscle but even with avidin–biotin amplification the LoD was restricted to $10 \, \mu g \, kg^{-1}$. The same antibody was incorporated into IAC columns, which were applied to the clean-up of CAP in pig muscle[97] and milk and eggs[98] prior to HPLC analysis. LoDs were well below $10 \, \mu g \, kg^{-1}$ in each matrix; for milk a 1 litre sample could be taken giving an LoD of $0.02 \, \mu g \, kg^{-1}$.

Nouws et al. have reported on the use of a commercially available test for CAP in which an ELISA assay is carried out employing antibody bound to a permeable membrane placed over an thin absorbent pad sheathed in plastic and resembling a credit card. Sample and reagents are drawn through the membrane into the pad. The test could be carried out on milk deproteinized with TCA[99] or on partially purified extracts from eggs[100] or tissue.[101] It was noted that there was significant batch-to-batch variation in the characteristics of the cards and their sensitivity was greater than that claimed by the manufacturer, which agrees with work done in this laboratory where in addition no false positives were found during the analysis of 108 milk samples.[102] Similarly, Laurensen and Nouws[103] found no false positives in a survey of CAP residues in muscle of 287 pigs, cattle, and sheep in a residue monitoring programme in the Netherlands, where the test was found to detect the presence of CAP at $1–3 \, \mu g \, kg^{-1}$.

2.8 Hormones and Other Steroids

The presence of small amounts of hormones has a profound influence upon animal growth characteristics and their use can markedly increase feed conversion and muscle formation. Diethylstilbestrol (DES) has been banned worldwide as a veterinary drug because of its carcinogenic and other adverse effects. The European Community has prohibited the use of other natural and xenobiotic hormones for fattening livestock, although the natural compounds are permitted for certain zootechnical and therapeutic purposes. When used in accordance with good veterinary practice, stilbene residues are not expected to exceed about $5 \, \mu g \, kg^{-1}$ in the liver and will be substantially less in other edible tissues.[104] Some steroidal hormones, such

[96] C. van de Water and N. Haagsma, J. Assoc. Offic. Anal. Chem., 1990, **73**, 534.

[97] C. van de Water, D. Tebbal, and N. Haagsma, J. Chromatogr., 1989, **478**, 205.

[98] C. van de Water and N. Haagsma, J. Chromatogr., 1987, **411**, 415.

[99] J. F. M. Nouws, J. Laurensen, and M. M. L. Aerts, Vet. Quart., 1988, **10**, 270.

[100] J. F. M. Nouws, J. Laurensen, and M. M. L. Aerts, Arch. Lebensmittelhyg., 1987, **38**, 7.

[101] J. F. M. Nouws, J. Laurensen, and M. M. L. Aerts, Arch. Lebensmittelhyg., 1987, **38**, 9.

[102] M. L. Bates, unpublished results.

[103] J. Laurensen and J. F. M. Nouws, Vet. Quart., 1990, **12**, 121.

[104] M. L. Bates, M. J. Warwick, G. Shearer, D. J. Harwood, I. D. Herriman, R. J. Heitzman, and D. H. Watson, J. Sci. Food. Agric., 1985, **36**, 31.

as melengestrol acetate, are, however, sufficiently lipophilic to be concentrated at higher levels in fat deposits. In order to detect hormone residues with any degree of certainty, LoDs of $0.1-0.5\ \mu g\,kg^{-1}$ are required. Under most circumstances this dictates extensive sample clean-up and the use of MS to achieve reliable evidence of treatment of animals with hormones. Methods are necessarily slow.

Hormones are described as oestrogens, gestagens, and androgens, depending upon their physiological action. Their structures are shown in Figure 7. One feature of the oestrogen group is the presence of a phenolic hydroxyl, and the pH-dependent partitioning of these compounds between organic and aqueous phases is frequently used for sample purification. This often alters the *cis:trans* ratio for DES residues and recovery estimates should be made on the basis of the sum of the two.

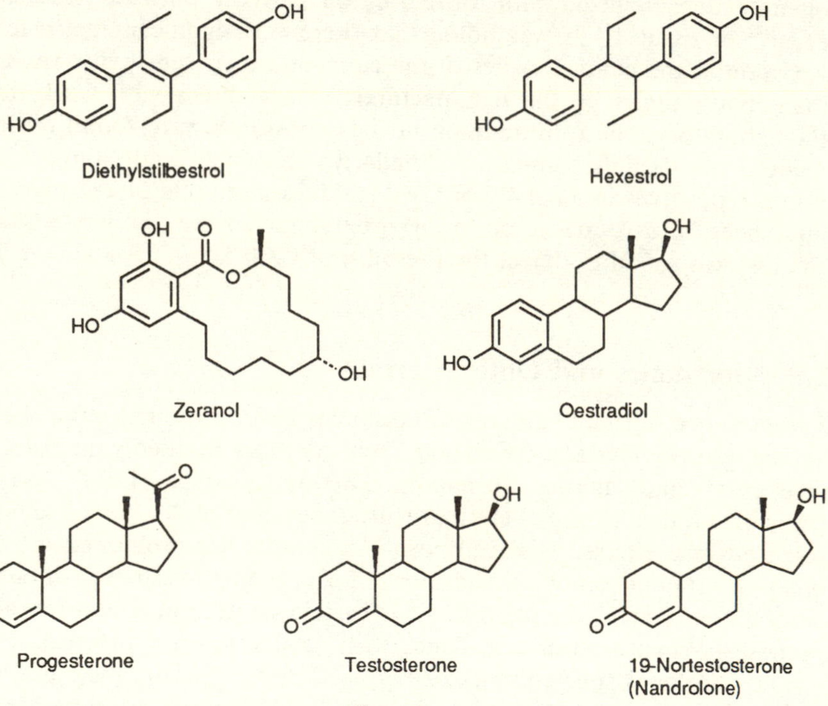

Diethylstilbestrol

Hexestrol

Zeranol

Oestradiol

Progesterone

Testosterone

19-Nortestosterone
(Nandrolone)

Figure 7

Hormones exhibit tissue-dependent conjugation to sulphates and glucuronides, metabolites which in liver account for most of any residue. In order to facilitate the analysis these products are converted back into the parent compounds by an initial hydrolysis. An enzyme extract from *Helix pomatia* is widely used for this purpose and is effective for the deconjugation of stilbenes but less so for some steroid hormones,[105] particularly 19-nortes-

---

[105] P. W. Tang and D. L. Crone, *Anal. Biochem.*, 1989, **182**, 289.

tosterone. Other aspects of the metabolism of DES, zeranol, trenbolone, and some steroids have been discussed by Metzler[106] together with methods of analysis for some of the metabolites.

Hormone residue analysis in foods was comprehensively reviewed by Ryan[107] in 1975. Procedures for zeranol[108] and trenbolone[109] have been reviewed more recently. Verbeke[110] developed a two-dimensional TLC method with LoDs of $1-10$ $\mu g\,kg^{-1}$ for most androgens and oestrogens, which worked well in this laboratory,[111] although it has been superseded by MS determination. After hydrolysis and defatting, hormones are partitioned into dichloromethane, cleaned up on XAD-2, and fractionated into oestrogens and androgens/gestagens on coupled alkaline Extrelut–alumina columns which are then eluted separately. A modified method with similar LoDs but claimed to be more efficient was recently described.[112]

Zeranol and its metabolites were determined by HPLC and electrochemical detection[113] with good sensitivity. Grob *et al.*[114] describe a coupled HPLC–GC method for DES in cattle urine although sensitivity was not adequate for tissue analysis. However, the technique has promise and could be useful for automated determination of other residues, particularly when employed in conjunction with a benchtop MS. In this laboratory a coupled-column size-exclusion–normal phase silica HPLC system[115] allowed UV detection of trenbolone at 360 nm with an LoD of *ca.* 1 $\mu g\,kg^{-1}$. Peak trapping and re-analysis on reverse phase HPLC gave an LoD of 0.1 $\mu g\,kg^{-1}$, primarily because of the better signal to noise ratio observed with this system.

Melengestrol acetate in cattle tissue has been analysed with an LoD of about 1 $\mu g\,kg^{-1}$ by GC–CI-MS after TLC clean-up of extracts.[116] Problems were experienced obtaining reproducible chromatography of the underivatized steroid. This compound has also been determined using a normal phase on-line HPLC column-switching combined clean-up and analytical system.[117] Extraction into acetonitrile was followed by addition of hexane and dichloromethane to give a three-phase system.[118] The middle phase was evaporated. Redissolved aliquots were injected on to a

[106] M. Metzler, *J. Chromatogr.*, 1989, **489**, 11.

[107] J. J. Ryan, *J. Chromatogr.*, 1975, **127**, 53.

[108] R. S. Baldwin, R. D. Williams, and M. K. Terry, *Regul. Toxicol. Pharmacol.*, 1983, **3**, 9.

[109] D. Schopper, *Fleischwirtschaft*, 1983, **63**, 406.

[110] R. Verbeke, *J. Chromatogr.*, 1979, **177**, 69.

[111] G. Shearer, M. Warwick, and M. L. Bates, unpublished results.

[112] L. van Look, Ph. Deschuytere, and C. van Peteghem, *J. Chromatogr.*, 1989, **489**, 213.

[113] J. E. Roybal, R. K. Munns, W. J. Morris, J. A. Hurlbut, and W. Shimoda, *J. Assoc. Offic. Anal. Chem.*, 1988, **71**, 263.

[114] K. Grob, H. P. Neukom, and R. Etter, *J. Chromatogr.*, 1986, **357**, 416.

[115] M. J. Shepherd and G. W. Stubbings, in preparation.

[116] E. E. Neidert, R. G. Gedir, L. J. Milward, C. D. Salisbury, N. P. Gurprasad, and P. W. Saschenbrecker, *J. Agric. Food Chem.*, 1990, **38**, 979.

[117] T. M. P. Chichila, P. O. Edlund, J. D. Henion, R. Wilson, and R. L. Epstein, *J. Chromatogr.*, 1989, **488**, 389.

[118] T. R. Covey, D. Silvestre, M. K. Hoffman, and J. D. Henion, *Biomed. Environ. Mass Spectrom.*, 1988, **15**, 45.

phenyl column and the appropriate fraction isolated on a silica concentration cartridge, which was eluted into the analytical silica column. UV detection was at 254 nm. The LoD for liver was 5 μg kg^{-1} and lower levels were quantitated in muscle. Column characteristics changed slowly in use; retention times were monitored and mobile phase composition altered appropriately. Confirmation was achieved by GC–MS of trapped HPLC fractions using heptafluorobutyrate derivatization.

A glucocorticosteroid, dexamethasone (Figure 8), was determined in cattle muscle and liver similarly.[119] Deconjugated tissue was extracted and cleaned up as described above except that the silica analytical column was replaced with a cyano packing. Different mobile phases were required for the separations. The dexamethasone peak was quantitated by UV at 239 nm and trapped for subsequent silylation and GC–EI-high resolution MS confirmation. An extraction efficiency of 70% (radiotracer experiments) was achieved at 8 μg kg^{-1}. Faster shaking was required for extraction from muscle, possibly because the matrix does not homogenize so finely as liver. LoDs were estimated as 4 and 6 μg kg^{-1} in muscle and liver, but accuracy and precision were not acceptable below 10 μg kg^{-1}. Betamethasone could not be distinguished from dexamethasone by either of the LC or GC–MS procedures.

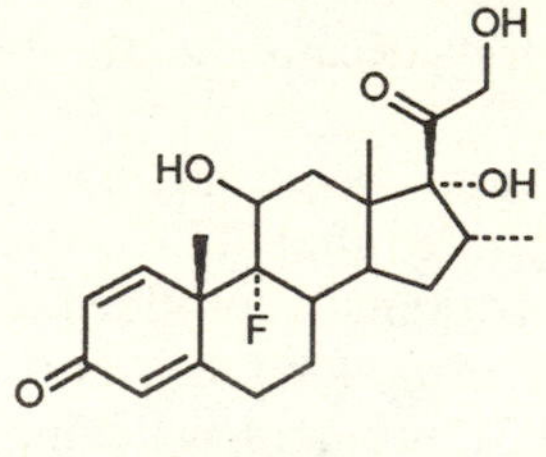

Dexamethasone

Figure 8

Shearan *et al.*[120] developed a method for dexamethasone in cattle tissues involving blending samples with 0.1 M sodium hydroxide, extraction into ethyl acetate, evaporation, defatting with hexane, C$_{18}$ SPE, and RP-HPLC with UV detection at 254 nm. An LoD of 10 μg kg^{-1} was reported.

Most of the methods used routinely for surveillance of hormones rely on MS to achieve adequate selectivity at the sub-μg kg^{-1} levels of interest. A GC–MS multi-hormone procedure developed by Bergner-Lang and Kächele[121] and based on the Verbeke[110] clean-up permits LoDs of

[119] L. G. McLaughlin and J. D. Henion, *J. Chromatogr.*, 1990, **529**, 1.
[120] P. Shearan, M. O'Keeffe, and M. R. Smyth, in Proc. Euroresidue Conf. Residue Vet. Drugs Food, ed. N. Haagsma, A. Ruiter, and P. B. Czedik-Eysenberg, University of Utrecht, Utrecht, 1990, p. 336.
[121] B. Bergner-Lang and M. Kächele, *Dtsch. Lebensm.-Rundsch.*, 1987, **83**, 349.

0.02–0.05 $\mu g\,kg^{-1}$ for both oestrogens and androgens and is stated to have been successfully collaboratively tested.[122] Although very lengthy, this is probably the method of choice for multi-hormone analysis. Direct MS–MS of extracts has been demonstrated[123] but to obtain the required selectivity with standard MS techniques necessitates careful sample clean-up, avoiding interferences arising from the system. For example, the hexestrol EI spectrum is restricted essentially to one ion at m/z 207 and some plastics can give rise to interferences. A GC–MS procedure for trenbolone permitting determination at 0.5 $\mu g\,kg^{-1}$ was reported by Hsu *et al.*[124]

The method employed in this laboratory for oestrogens (DES, hexestrol, dienestrol, oestradiol, and zeranol) is based on the three-phase partition procedure of Covey *et al.*[118] The middle phase is taken through further clean-up, including anion exchange chromatography. In our hands the ion-exchange column often became overloaded, so a modified protocol was devised.[125]

In outline, liver is homogenized in pH 4.1 buffer and deuterated internal standards are added as appropriate. After *Helix pomatia* hydrolysis, the pH is adjusted to 9.2 and the digest subjected to solid-phase partition on an Extrelut column with 5% pentan-2-ol in hexane. The eluant is partitioned against base, with the addition of methanol to avoid emulsions. The aqueous layer is cleaned up on a resin cartridge under sequential anion exchange/reverse phase conditions. To obtain adequate recoveries of zeranol on the final elution with methanol, 0.01% acetic acid must be added to suppress ionization of this oestrogen. However, if the concentration of acetic acid is raised above 0.01%, a reaction involving liver co-extractives (standards give no problems) reduces the recovery of hexestrol. For GC–MS analysis, the residue is evaporated and silylated with BSTFA containing 2% trimethylsilylimidazole.

It is necessary to adopt strict procedures to avoid the possibility of cross-contamination giving false positive results. All concentrated standards must be segregated and the glassware kept separate from that used for analysis of tissue samples. Glassware which has come into contact with positive samples should be cleaned with chromic acid. Injections of standards on capillary GC should be followed by blank silylating reagent before analysis of samples as carry-over has been observed.

Immunoassay has been used for many years for hormone analysis, and has recently been reviewed.[126] RIA and EIA immunoassay kits may be obtained from a number of manufacturers (see Table 4). One RIA developed for the analysis of stilbenes in cattle liver was evaluated in this laboratory.[127] Sample preparation required about 60% of the time taken

[122] B. Bergner-Lang and M. Kächele, *Dtsch. Lebensm.-Rundsch.*, 1989, **85**, 78.
[123] R. Maffei Facino, M. Carini, A. Da Forno, P. Traldi, and G. Pompa, *Biomed. Environ. Mass Spectrom.*, 1986, **13**, 121.
[124] S.-H. Hsu, R. H. Eckerlin, and J. D. Henion, *J. Chromatogr.*, 1988, **424**, 219.
[125] G. W. Stubbings and M. J. Shepherd, in preparation.
[126] H. H. D. Meyer, *Arch. Lebensmittelhyg.*, 1990, **41**, 4, 6–7.
[127] J. M. Carter and M. J. Shepherd, unpublished results.

for the MS method described above. The LoD was calculated from the standard curve using the mean plus two standard deviations of six determinations of a blank tissue. The values found were: cattle liver 0.74 μg kg^{-1}, pig liver 0.55 μg kg^{-1}, turkey muscle 0.50 μg kg^{-1}, chicken muscle 0.41 μg kg^{-1}, and duck muscle 0.38 μg kg^{-1}. If the error of plotting the standard curve were taken into account, these figures would increase significantly. In addition, the blanks all fell close to the high end of the calibration curve. From these results the kit was judged unsuitable for use in national food surveillance.

There is no information available to evaluate the performance of this or similar products for other applications. In general hormone immunoassays for residues in tissue, particularly where one assay is required to accept different tissues from a range of species, are liable to false positives unless great care is taken over sample clean-up. HPLC purification of extracts is widely used.[126]

An early use[18] of immunoaffinity clean-up was to concentrate oestradiol from serum prior to MS analysis. More recently, it has been employed for the isolation of methyltestosterone[128] and 19-nortestosterone[128,129] for GC–MS determination. Tissues were proteolysed in tris buffer, hormones were partitioned into methanol and defatted,[128] or adsorbed onto XAD-2 and eluted with methanol,[129] and the residue was cleaned up by IAC for GC–MS. Very low LoDs may be attained by the use of IAC; 0.05 μg kg^{-1} was reported for 19-nortestosterone.[129] Bagnati *et al.*[130] determined stilbenes in biological fluids at the 0.01 μg kg^{-1} level via IAC clean-up, pentafluorobenzyl derivatization, and GC–negative ion CI-MS. The columns used were supplied by Genego. An interesting application[131] of IAC to the analysis of cattle urine was as a precolumn in a fully automated HPLC system. Experience within this laboratory of in-house and commercial trenbolone IAC columns has been encouraging, with an LoD of 0.25 μg kg^{-1} in cattle liver using direct HPLC–UV analysis of analytes as desorbed from the columns with aqueous alcohol.[132]

Other biologically based assays have been reported which detect oestrogens selectively by their interaction with uterine receptors.[133] The individual drugs were separated by HPLC before quantitation in a competition assay using radiolabels. The hormone receptors required are not commercially available.

[128] L. A. van Ginkel, R. W. Stephany, H. J. van Rossum, H. M. Steinbuch, G. Zomer, E. van de Heeft, and A. P. J. M. de Jong, *J. Chromatogr.*, 1989, **489**, 111.

[129] W. Haasnoot, R. Schilt, A. R. M. Hamers, F. A. Huf, A. Farjam, R. W. Frei, and U. A. Th. Brinkman, *J. Chromatogr.*, 1989, **489**, 157.

[130] R. Bagnati, M. Grazia Castelli, L. Airoldi, M. Paleologo Oriundi, A. Ubaldi, and R. Fanelli, *J. Chromatogr.*, 1990, **527**, 267.

[131] A. Farjam, G. J. de Jong, R. W. Frei, U. A. Th. Brinkman, W. Haasnoot, A. R. M. Hamers, R. Schilt, and F. A. Huf, *J. Chromatogr.*, 1988, **452**, 419.

[132] M. J. Shepherd, M. Pointer, and J. M. Carter, in preparation.

[133] C. J. M. Arts and H. van den Berg, *J. Chromatogr.*, 1989, **489**, 225.

2.9 Ionophore Polyethers

Weiss and MacDonald[134] have discussed the properties of ionophore polyether antibiotics and methods for their analysis. The many polar sites within the drug molecules (see Figure 9) readily interact with and enclose metal ions, leaving a lipophilic surface for solvation. Those drugs containing spiroketal structures, including monensin, are acid labile, while those containing β-hydroxy ketone groups are base and heat labile. Most of these drugs have no useful chromophore but lasalocid contains a salicylate

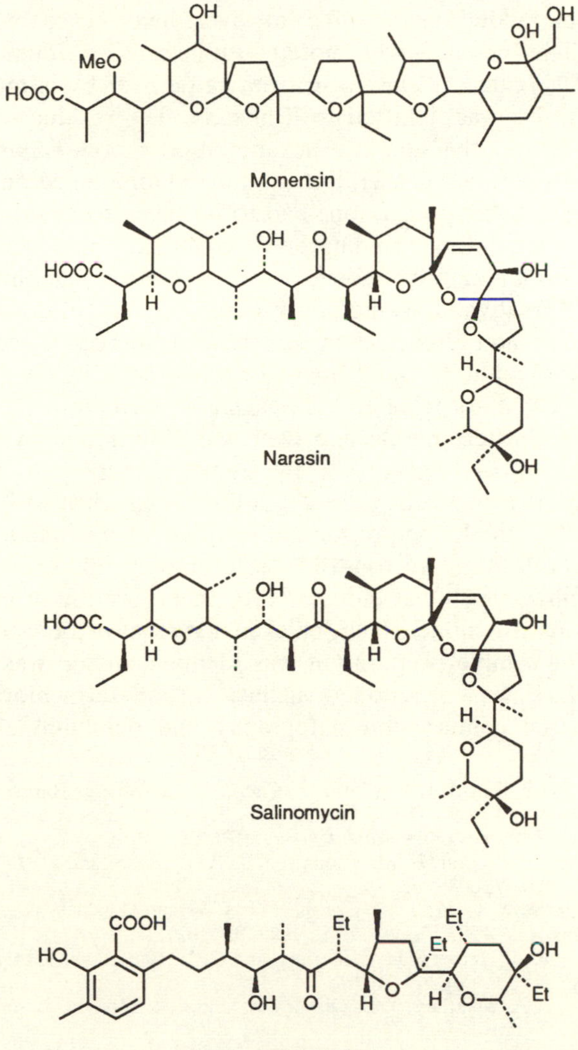

Monensin

Narasin

Salinomycin

Lasalocid A

Figure 9

[134] G. Weiss and A. MacDonald, *J. Assoc. Offic. Anal. Chem.*, 1985, **68**, 971.

moiety and is fluorescent, particularly in the ionized form. The intensity of fluorescence increases 100-fold[135] between pH 3.2 and 8.3. The excitation and emission spectra of this compound (solvent unspecified) have been presented.[134,136]

The ionophores are extensively metabolized and unchanged drug accounts for only a minor proportion of the total residue in liver.[137,138] Potential problems with co-extracted carotenoid interferences have been noted.[134,136] Fat is the target tissue; concentrations of metabolites are highest in the liver.[139,140] Residues were stable in liver at $-20\,°C$ for at least one year.[135] More than 15% loss of lasalocid over 1 hour was reported when added as a spike to liver held at room temperature; conversely, slight increases were noted with incurred residues.[134]

Weiss *et al.*[135] extracted lasalocid from cattle liver by homogenizing with acetonitrile, which was defatted with hexane. The residue was partitioned between water and the alkaline hexane-based mobile phase, which was injected directly into a silica HPLC column. Fluorescence detection was at 310 and 430 nm. Strict precautions had to be taken to avoid cross-contamination of samples during grinding and extraction. The LoD was given as $25\ \mu g\,kg^{-1}$; chromatograms shown in an accompanying paper[141] indicate this to be conservative. This method was subjected to interlaboratory study for cattle liver[142] and also chicken skin[143] with, in both cases, encouraging results. Results could be confirmed at levels well below $25\ \mu g\,kg^{-1}$ by pyrolysis GC–CI-MS of trapped LC peaks after silylation.[141]

Lasalocid in chicken muscle and liver was determined by Horii *et al.*[144] with an LoD of *ca.* $10\ \mu g\,kg^{-1}$ by homogenizing with methanol and partitioning into carbon tetrachloride, followed by silica SPE clean-up and RP-HPLC in a mobile phase containing pH 7.0 phosphate buffer, with fluorescence detection. Acetonitrile–buffer mobile phases were found to cause peak tailing which was eliminated by incorporation of methanol.

Two methods for monensin in chicken tissue have been collaboratively studied.[145] The sample preparation was identical; tissue was blended with methanol, which was partitioned against carbon tetrachloride. Clean-up was by silica gel column chromatography and detection either by TLC

[135] G. Weiss, N. R. Felicito, M. Kaykaty, G. Chen, A. Caruso, E. Hargroves, C. Crowley, and A. MacDonald, *J. Agric. Food Chem.*, 1983, **31**, 75.

[136] A. MacDonald, *J. Assoc. Offic. Anal. Chem.*, 1978, **61**, 1214.

[137] G. P. Dimenna, F. S. Lyon, F. M. Thompson, J. A. Creegan, and G. J. Wright, *J. Agric. Food Chem.*, 1989, **37**, 668.

[138] G. P. Dimenna, F. S. Lyon, J. A. Creegan, G. J. Wright, L. C. Wilkes, D. E. Johnson, and T. Szymanski, *J. Agric. Food Chem.*, 1990, **38**, 1029.

[139] A. L. Donoho, R. J. Herberg, L. L. Zornes, and R. L. Van Duyn, *J. Agric. Food Chem.*, 1982, **30**, 909.

[140] A. L. Donoho, J. A. Manthey, J. L. Occolowitz, and L. L. Zornes, *J. Agric. Food Chem.*, 1978, **26**, 1090.

[141] G. Weiss, M. Kaykaty, and B. Miwa, *J. Agric. Food Chem.*, 1983, **31**, 78.

[142] D. R. Newkirk and C. J. Barnes, *J. Assoc. Offic. Anal. Chem.*, 1989, **72**, 581.

[143] L. R. Frank and C. J. Barnes, *J. Assoc. Offic. Anal. Chem.*, 1989, **72**, 584.

[144] S. Horii, K. Miyahara, and C. Momma, *J. Liq. Chromatogr.*, 1990, **13**, 1411.

[145] Analytical Methods Committee, *Analyst (London)*, 1986, **111**, 1089.

(97 + 3, ethyl acetate–water) with bioautography detection or by TLC visualized after spraying with acidic anisaldehyde. The methods were considered equally effective, with variable recovery and LoDs of $100 \, \mu g \, kg^{-1}$.

In this laboratory a similar procedure is used to analyse lasalocid, monensin, narasin, and salinomycin simultaneously in chicken muscle and egg at levels down to $10 \, \mu g \, kg^{-1}$: acetonitrile extraction, addition of saline, partition into carbon tetrachloride, silica SPE, and HPTLC. The native fluorescence of lasalocid is seen under UV irradiation of the wet plate; it fades rapidly as the solvent evaporates. The other ionophores are visualized with the anisaldehyde reagent.[146]

Monensin was determined by HPLC of a fluorescent derivative with an LoD of $1 \, \mu g \, kg^{-1}$ in chicken and cattle tissue,[147] although the procedure seems rather laborious. The synthesis of 9-anthryldiazomethane (ADAM) is required, which, however, was reported to be stable in ether solution at $-20 \, °C$ for at least 5 months.

ADAM was also employed in a complex multi-residue method[148] for monensin, salinomycin, and narasin in cattle liver. Tissue was extracted with methanol–water and purification performed with alumina and Sephadex LH-20 column chromatography. Lasalocid was taken through a parallel procedure and treated with ADAM; other ionophores were acetylated before preparation of the fluorescent derivatives. LoDs of $150 \, \mu g \, kg^{-1}$ were established. No estimate of sample throughput was given but the method appears slow.

Salinomycin in chicken skin and fat was determined[137] using narasin as internal standard. Samples were extracted with iso-octane which was purified by silica SPE. Ionophores were oxidized with pyridinium dichromate and the products purified by silica SPE and separated by RP-HPLC with column switching. Despite salinomycin oxidation, the UV detection wavelength was only 225 nm. Lasalocid and monensin gave no response.[149] Recoveries were excellent and the LoD was considered to be $5 \, \mu g \, kg^{-1}$ for liver and $20 \, \mu g \, kg^{-1}$ for skin and fat.[137]

FAB-MS was used for analysis of monensin A extracted from chicken fat and purified by TLC. Monensin B was employed as internal standard. An LoD of under $10 \, \mu g \, kg^{-1}$ was obtained but confirmation was not adequately established because FAB-MS gives limited fragmentation of monensin.[150] Maduramicin-α was confirmed in chicken fat at levels of about $500 \, \mu g \, kg^{-1}$ using thermospray LC–MS-MS.[151]

146 J. Tarbin and W. H. H. Farrington, unpublished results.

147. K. Takatsuki, S. Suzuki, and I. Ushizawa, *J. Assoc. Offic. Anal. Chem.*, 1986, **69**, 443.

148 E. E. Martinez and W. Shimoda, *J. Assoc. Offic. Anal. Chem.*, 1986, **69**, 637.

149 G. P. Dimenna, J. A. Creegan, L. B. Turnbull, and G. J. Wright, *J. Agric. Food Chem.*, 1986, **34**, 805.

150 G. B. Blomkvist, K. M. Jansson, E. R. Ryhage, and B.-G. Osterdahl, *J. Agric. Food Chem.*, 1986, **34**, 274.

151 S. J. Stout, L. A. Wilson, A. I. Kleiner, A. R. daCunha, and T. J. Francl, *Biomed. Environ. Mass Spectrom.*, 1989, **18**, 57.

2.10   Macrocyclic lactones

2.10.1   *Ivermectin*

Ivermectin (IVM, see Figure 10) is a mixture of closely related drugs but is required to contain at least 80% of the B_{1a} compound, which is sometimes referred to as avermectin H_2B_{1a} because IVM is produced by reduction of a double bond in the fermentation product avermectin. IVM B_{1a} is metabolized more slowly than the other drug components and so occurs as the major unaltered residue. The target tissues are fat and liver. In cattle, the drug persists longest in liver, with a half-life of 4.9 days.[152] Demethylated and monosaccharide metabolites occur.[153] The drug may be monitored by UV, with an absorption maximum in methanol of 244 nm, but detection is often carried out after induction of fluorescence by dehydration.

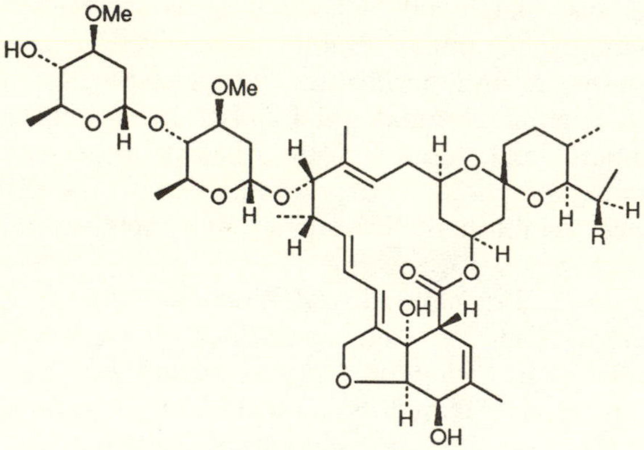

Ivermectin B_{1a} R = Et
B_{1b} R = Me

Figure 10

A method employing fluorescence detection for IVM in cattle and sheep tissue and fat was devised by Tway *et al.*[152] Tissue was extracted into iso-octane and the residue subjected to extensive liquid–liquid partitioning, dehydrated with acetic anhydride in the presence of methylimidazole to induce fluorescence, and purified by silica SPE. The lowest concentration at which reproducible results were obtained was 10 μg kg^{-1}, with recoveries of *ca.* 80%. The procedure is slow, with one analyst capable of determining 12 samples in two days. Results from radiolabelled incurred residues were in close agreement with those obtained with an isotope

[152] P. C. Tway, J. S. Wood, and G. V. Downing, *J. Agric. Food Chem.*, 1981, **29**, 1059.
[153] S.-H. L. Chiu, E. Sestokas, and R. Taub, *J. Chromatogr.*, 1988, **433**, 217.

dilution assay. The method is effective for cooked cattle and pig muscle.[154]

Confirmation of IVM in cattle liver was attempted by a number of MS techniques.[155] Direct insertion ammonia CI-MS–MS was the only method found satisfactory, despite protracted sample clean-up, and a limit of detection of 5–10 $\mu g\, kg^{-1}$ was estimated. Residues were isolated as above and purified by Florisil column chromatography and preparative RP-HPLC.

Radiolabelled IVM was separated from pig metabolites by normal phase HPLC with an iso-octane–methanol mobile phase,[153] and the metabolite fraction rechromatographed on RP-HPLC using acetonitrile–methanol–water. Each of the metabolites was susceptible to fluorescence detection after dehydration.

Recently, two rapid methods for IVM have been reported. Nordlander and Johnsson[156] extracted pig muscle with acetonitrile. This was diluted with water and triethylamine added to block unmodified silanols during clean-up on a C_8 SPE cartridge. The residue was then dehydrated, purified by silica SPE, and analysed by RP-HPLC in methanol–water (97:3) with fluorescence detection at excitation and emission wavelengths of 364 and 470 nm. The LoD was *ca.* 1 $\mu g\, kg^{-1}$ and throughput was stated to be 50 samples per analyst per week. Problems were experienced with interferences originating in some batches of the silica SPE cartridges. Correlation with the original Tway *et al.*[152] method was good. The method has been used successfully in this laboratory.

Dickinson[157] employed the fluorescent dehydration product of IVM as an internal standard for a sample preparation method involving extraction into aqueous acetonitrile, deproteinization with zinc sulphate, and C_{18} SPE. An LoD of 2 $\mu g\, kg^{-1}$ was achieved but the method was not acceptable for liver.

2.10.2 Macrolide Antibiotics

These drugs are typically complexes of several closely related compounds which may vary in proportion depending upon the source. The structures of some of these compounds are shown in Figure 11. Effective monitoring of the macrolides will require the development of more rapid and sensitive methods than are currently available. Moats[158] has reviewed earlier TLC and HPLC methodology for the determination of tylosin (TY) and erythromycin (EM).

Macrolides are extensively metabolized.[159] TY is rapidly cleared *in vivo* and there is evidence that residues are slowly lost on frozen storage.[158]

[154] P. Slanina, J. Kuivinen, C. Ohlsen, and L.-G. Ekstrom, *Food Add. Contam.*, 1989, **6**, 475.
[155] P. C. Tway, G. V. Downing, J. R. B. Slayback, G. S. Rahn, and R. K. Isensee, *Biomed. Mass. Spectrom.*, 1984, **11**, 172.
[156] I. Nordlander and H. Johnsson, *Food Add. Contam.*, 1990, **7**, 79.
[157] C. M. Dickinson, *J. Chromatogr.*, 1990, **528**, 250.
[158] W. A. Moats, *J. Assoc. Offic. Anal. Chem.*, 1985, **68**, 980.
[159] J. Okada and S. Kondo, *J. Assoc. Offic. Anal. Chem.*, 1987, **70**, 818.

Macrolides other than oleandomycin (OM) are stated to be unstable in acidic or basic aqueous solutions; manipulations under these conditions should be carried out rapidly.[160] Spiramycin component I (SM I) was unstable in spiked tissue during clean-up.[161] Liver was shown to be the target organ. TY has a UV absorbance at 280 nm, but both EM and TY are more readily detected electrochemically.

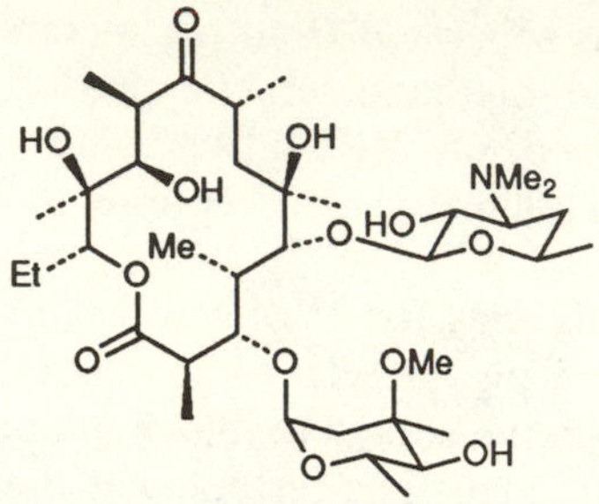

Erythromycin A

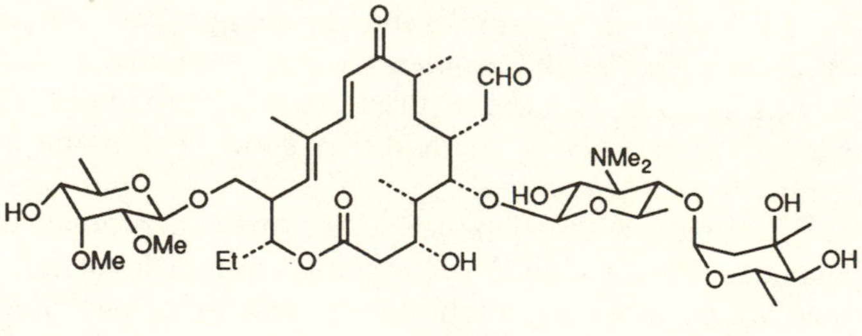

Tylosin

Figure 11

Moats devised a method[158] for TY involving extraction of milk, eggs, or muscle with aqueous acetonitrile. Liver and kidney required the use of pH 2.2 buffer to give adequate recoveries. TY was partitioned into dichloromethane and analysed on an RP-HPLC column in a pH 4.6 phosphate-buffered mobile phase, although evidence was presented which indicates that the chromatographic separation was taking place on residual silanols. The LoD was considered to be below 100 μg kg^{-1} for milk and at that level for tissue.

TY, EM, OM, and spiramycin (SM) in tissue, egg, and milk[160] were extracted with acetonitrile at pH 8.5 and subjected to acid–base partition. Macrolides were separated by TLC in n-butanol–water–acetic acid

[160] M. Petz, R. Solly, M. Lymburn, and M. H. Clear, *J. Assoc. Offic. Anal. Chem.*, 1987, **70**, 691.
[161] P. Sanders, P. Guillot, G. Moulin, B. Delepine, and D. Mourot, in Proc. Euroresidue Conf. Residue Vet. Drugs Food, ed. N. Haagsma, A. Ruiter, and P. B. Czedik-Eysenberg, University of Utrecht, Utrecht, 1990, p. 315.

(6 + 2 + 2) and visualized with acidified xanthydrol, which yielded purple spots and an LoD of 20 $\mu g\,kg^{-1}$. Recoveries were 50–80%. The reported R_f for SM appears to be in error by comparison with the photograph of a developed TLC plate. Individual alternative reagents are described for partial confirmation of TY and EM, together with considerable information on alternative derivatization procedures. Different chromatographic conditions were required to distinguish TY from a potential interference. Electrochemical detection was insufficiently selective to permit quantitation by RP-HPLC.

SM I and its metabolite neospiramycin were extracted[161] from cattle tissues with chloroform in the presence of phosphate buffer, cleaned up by silica SPE, and injected directly from the cartridges into a RP-HPLC column with acetonitrile–0.5% sulphuric acid eluant and UV detection at 231 nm. LoDs were 50 IU kg^{-1} for SM I and 50 $\mu g\,kg^{-1}$ for neospiramycin. Degradation of SM I during extraction and clean-up was prevented by addition of cobalt chloride. A C_8-RP-HPLC method[162] is available for SM in chicken tissues at a stated LoD of 50 $\mu g\,kg^{-1}$.

Sedecamycin and three metabolites were determined in pig tissues and fat[159] by blending in pH 4.5 buffer and extracting into ethyl acetate, followed by combined Florisil–silica clean-up, which was shown to be a critical step in the analysis. Analysis was by normal phase HPLC with detection at 227 nm. Recoveries were 70–90% and LoDs were in the range 12–44 $\mu g\,kg^{-1}$. The method was thoroughly evaluated and analyte water–solvent partition constants are presented. Horie *et al.* have used RP-HPLC with UV detection to separate the components of the TY complex after solvent extraction from cattle, chicken, or pig tissue. An LoD of 50 $\mu g\,kg^{-1}$ was reported.[163]

A GC–EI-MS method for EM in cattle and pig muscle was developed[164] by Takatsuki *et al.* After extraction with methanol, defatting, base partition, and silica gel clean-up, EM was hydrolysed and acetylated. Single ion monitoring at m/z 200 was employed because this fragment dominates the EI spectrum. An LoD of 10 $\mu g\,kg^{-1}$ was estimated and full spectrum confirmation could be obtained at EM concentrations in excess of 1000 $\mu g\,kg^{-1}$.

2.11 Nitro-drugs

2.11.1 Nitrofurans and Other Nitro Drugs Except Nitroimidazoles

Nitrofurans (Figure 12) degrade readily and dilute standard solutions

[162] T. Nagata and M. Saeki, *J. Assoc. Offic. Anal. Chem.*, 1986, **69**, 644.
[163] M. Horie, K. Saito, Y. Hoshino, N. Nose, and H. Nakazawa, *Eisei Kagaku*, 1988, **34**, 128 (*Chem. Abstr.*, 1988, **109**, 169045).
[164] K. Takatsuki, S. Suzuki, N. Sato, and I. Ushizawa, *J. Assoc. Offic. Anal. Chem.*, 1987, **70**, 708.

should be prepared daily. They are also susceptible to photolysis, particularly by sunlight, and manipulations must be carried out only under the minimum of incandescent lighting. In early procedures silylation of glassware was recommended, but this seems not to be necessary. The drugs have UV absorption maxima around 360 nm, providing good sensitivity and selectivity for HPLC detection. They also give a good response to the electrochemical detector. Several multi-residue methods incorporate nitrofurans but with inadequate LoDs apart from the Malisch procedure[35] which provides borderline detectability.

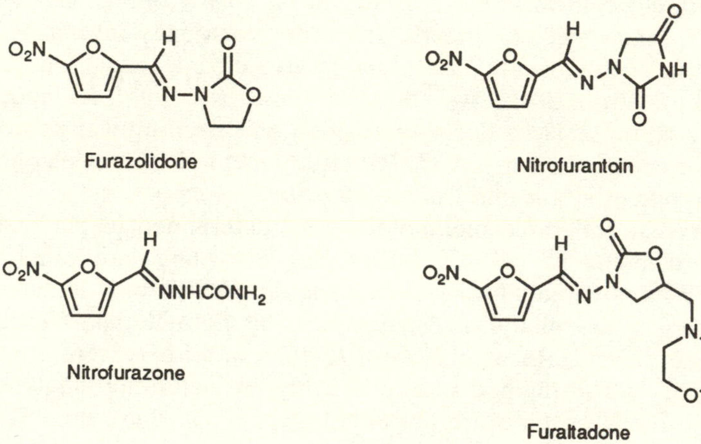

Furazolidone

Nitrofurantoin

Nitrofurazone

Furaltadone

Figure 12

Nitrofurans are rapidly metabolized in animals. Furazolidone (FZD), for example, was present at levels of less than $2 \mu g \, kg^{-1}$ in muscle, liver, and fat only 2 hours after withdrawal from pigs[165] although a high concentration of metabolites was evident from radiotracer experiments. Metabolites, both free and protein-bound, were three-fold higher in pig liver and kidney compared with muscle. However, metabolism is thought not to continue in the developing egg.[166] An RP-HPLC method involving acid hydrolysis and derivatization of the cleavage product has been demonstrated[167] for determination of part of the bound residues arising from FZD–protein interactions *in vitro*.

Incurred residue in pig muscle was almost completely lost[168] after storage for only two weeks at $-30\,°C$, but microwaving samples before

[165] L. H. M. Vroomen, M. C. J. Berghmans, P. van Leeuwen, T. D. B. van der Struijs, P. H. U. de Vries, and H. A. Kuiper, *Food Add. Contam.*, 1986, **3**, 331.

[166] N. A. Botsoglou, *J. Agric. Food Chem.*, 1988, **36**, 1224.

[167] L. A. P. Hoogenboom, M. van Kammen, M. C. J. Berghmans, and H. A. Kuiper, in Proc. Euroresidue Conf. Residue Vet. Drugs Food, ed. N. Haagsma, A. Ruiter, and P. B. Czedik-Eysenberg, University of Utrecht, Utrecht, 1990, p. 221.

[168] G. Carignan, A. I. MacIntosh, and S. Sved, *J. Agric. Food Chem.*, 1990, **38**, 716.

freezing considerably improved nitrofuran stability. Degradation of residues could be prevented[169] by freezing tissue after homogenizing with two volumes of 1.5 M potassium dihydrogen phosphate solution containing 0.2% sodium azide. Samples were wrapped in aluminium foil and stored at −20 °C.

Analysis of nitrofurans typically consists of extraction into dichloromethane, defatting, and RP-HPLC analysis with a UV detector set at 365 nm. Laurensen and Nouws[169] extracted FZD, furaltadone (FAD), nitrofurantoin (NFT), and nitrofurazone (NFZ) from stabilized tissue homogenates by addition of acetonitrile and partition into dichloromethane–ethyl acetate. Analysis was by cyano RP-HPLC and UV at 365 nm and LoDs were 1 $\mu g\,kg^{-1}$. A range of other drugs were examined for interference; carbadox eluted just before NFT.

Aerts *et al.*[11] employed dialysis for determination of FZD, NFZ, FAD, and NFT in milk, eggs and muscle. Milk was defatted by centrifugation and diluted 1:1 with saline prior to dialysis. Egg homogenate was similarly diluted. Muscle was blended with saline and the supernatant injected. RP-HPLC in a pH 5 buffered mobile phase afforded LoDs (UV, 365 nm) for muscle and egg of 1–2 $\mu g\,kg^{-1}$ for FZD, NFT, and NFZ and 3–5 $\mu g\,kg^{-1}$ for FAD. NFT and NFZ in milk could not be quantitated because of interference, but FAD and FZD LoDs were 10 and 5 $\mu g\,kg^{-1}$. The method was rapid, permitting analysis of up to 30 samples per day. It merits close attention, particularly if it could be developed to improve sensitivity and detect metabolites.

Parks[170] has described an electrochemical (EC) procedure for FZD, NFZ, sulphanitran, and three related compounds, aklomide, dinitolmide, and nitromide (see Figure 13). Chicken muscle or liver was extracted with chloroform–ethyl acetate–dimethyl sulphoxide (DMSO) and cleaned up on neutral alumina. Analysis was carried out by RP-HPLC in pH 6 buffered mobile phase containing EDTA with an EC detector set at −0.8 V against Ag/AgCl. Sample throughput was eight per analyst per day. It was stated that the column required 5–7 days conditioning with mobile phase to avoid excessive peak tailing of the nitrofurans, which is not generally found to be a problem. Only partial separation was achieved of FZD and NFZ. LoDs

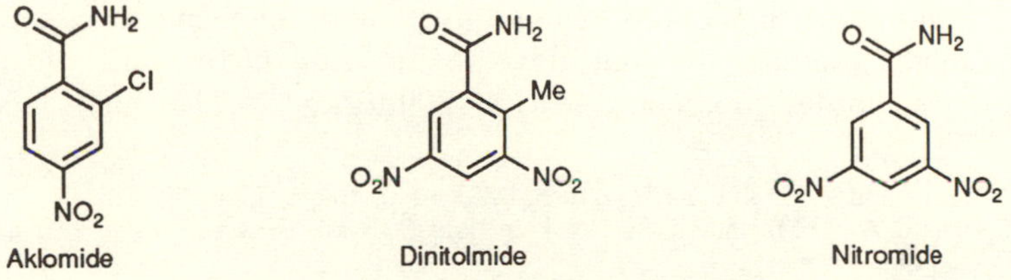

Aklomide   Dinitolmide   Nitromide

Figure 13

[169] J. J. Laurensen and J. F. M. Nouws, *J. Chromatogr.*, 1989, **472**, 321.
[170] O. W. Parks, *J. Assoc. Offic. Anal. Chem.*, 1989, **72**, 567.

were: nitrofurans 2.5 μg kg^{-1}; aklomide, nitromide and dinitolmide 2 μg kg^{-1}; sulphanitran 6 μg kg^{-1}.

An interference to NFZ was found to accumulate in frozen liver. Peak subtraction by photochemical reaction of the nitrofurans was suggested as a partial confirmation procedure. Conditions are given for resolution of the products. In a subsequent publication, Parks and Kubena[171] describe the application of this method to the analysis of a longer-lived metabolite of FZD. Another method employing EC detection for dimetridazole[172] was found to be effective with minor modification for the determination of FZD in pig muscle, liver, and kidney.[168] An LoD of 0.2 μg kg^{-1} was achieved.

FZD in turkey fat, liver, kidney, muscle, or skin[173] was extracted with dichloromethane and the residue subjected to extensive liquid–liquid partitioning before RP-HPLC with a pH 5 mobile phase and UV detection at 365 nm. Additional clean-up was carried out if necessary by C$_{18}$ SPE. Recoveries were 75–90% at 0.5 μg kg^{-1}. FZD in egg was determined in a similar manner by acidification to pH 4 and extraction with dichloromethane.[166] The residue, in acetone, was cooled to −78 °C and filtered (to avoid blockage of HPLC column frits), before liquid–liquid partition. Partial confirmation of positive results could be obtained either by diode array peak characterization or by collecting the peak and carrying out silica TLC, visualizing FZD by production of a fluorescent product through reaction with pyridine. Interferences which originated from contact of filter papers with the acidic extractant were eliminated by use of sintered glass funnels. Recovery of FZD was excellent at concentrations of *ca*. 5 μg kg^{-1} and the LoD was below 1 μg kg^{-1}.

The MSPD procedure[12] has been applied to the determination of FZD in milk,[174] and chicken tissue[175] at levels down to 8 μg kg^{-1}. Prepared matrix is washed with hexane and eluted with dichloromethane. The method potentially has advantages for rapid screening. FZD in salmon was determined by Samuelson[176] by extraction into acid-buffered methanol and partition into dichloromethane followed by amino SPE clean-up. UV detection at 400 nm was employed to eliminate an interference. An LoD of 5 μg kg^{-1} was reported.

The 4,4-dinitrocarbanilide (DNC) component of nicarbazin (Figure 14) was determined in chicken muscle and liver by Parks using an electrochemical method[177] similar to that developed[170] for nitrofurans. A reagent interference limited detection to 100–200 μg kg^{-1}. The high recoveries

[171] O. W. Parks and L. F. Kubena, *J. Assoc. Offic. Anal. Chem.* 1990, **73**, 526.

[172] G. Carignan, A. I. Macintosh, W. Skakum, and S. Sved, *J. Assoc. Offic. Anal. Chem.*, 1988, **71**, 1146.

[173] W. Winterlin, G. Hall, and C. Mourer, *J. Assoc. Offic. Anal. Chem.*, 1981, **64**, 1055.

[174] A. R. Long, L. C. Hsieh, M. S. Malbrough, C. R. Short, and S. A. Barker, *J. Agric. Food Chem.*, 1990, **38**, 430.

[175] M. M. Soliman, A. R. Long, and S. A. Barker, *J. Chromatogr.*, 1990, **13**, 3327.

[176] O. B. Samuelsen, *J. Chromatogr.*, 1990, **528**, 495.

[177] O. W. Parks, *J. Assoc. Offic. Anal. Chem.*, 1988, **71**, 778.

achieved (97–105% at 250–8000 μg kg^{-1}) were attributed to the inclusion of DMSO in the extraction solvent to prevent tissue binding. Another method[178] for DNC in chicken tissue, fat, and skin involved extraction with ethyl acetate, defatting with hexane, and alumina clean-up. Analysis was by RP-HPLC using methanol–water (3 + 1) and UV detection at 340 nm. The LoD was 20 μg kg^{-1}, with recoveries at 100 μg kg^{-1} of 80–85%.

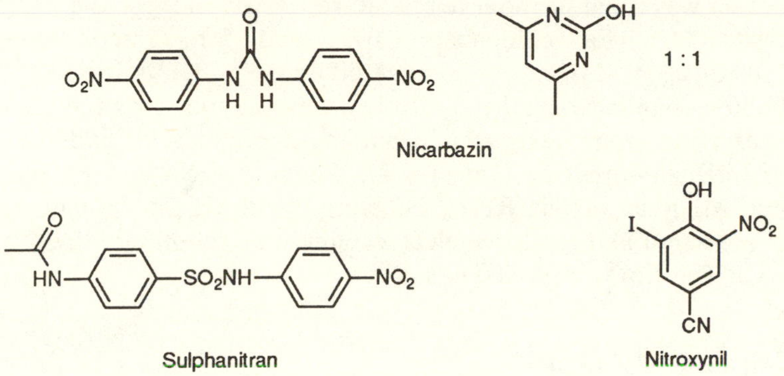

Nicarbazin

Sulphanitran

Nitroxynil

Figure 14

Confirmation was achieved by thermospray LC–MS. It was stated that DNC tends to adsorb to glassware; care should be taken to avoid cross-contamination and final residues should not be evaporated to complete dryness. Standards and extracts were stable in mobile phase under refrigeration for at least one month; residues were stable in frozen chicken tissue for at least 10 months. Nicarbazin in eggs[179] was extracted with acetonitrile in the presence of acetic acid, water removed with sodium sulphate, and residues cleaned up by silica SPE. Analysis on RP-HPLC in acetonitrile–pH 4.8 acetate buffer with UV detection at 360 nm gave a recovery of 90% at 5–25 μg kg^{-1} with an LoD of 2.5 μg kg^{-1}.

Thermospray LC–MS was also used for nitroxynil[180] (Figure 14) in cattle and sheep tissue. Samples were blended with basic phosphate buffer, acidified, and extracted with ether, then back-extracted into the phosphate buffer, and the partition cycle was repeated. Residues were separated by RP-HPLC in acetonitrile-0.1 M ammonium acetate and detected by negative ion MS. Only two ions were available for monitoring. Recovery at 10 μg kg^{-1} was 84% and the LoD was 2 μg kg^{-1}. Nitroxynil in milk[181] was isolated by acid–base partition, derivatized with diazomethane, and determined by packed column GC–ECD. The LoD was 10 μg kg^{-1}.

[178] J. L. Lewis, T. D. Macy, and D. A. Garteiz, *J. Assoc. Offic. Anal. Chem.*, 1989, **72**, 577.
[179] M. H. Vertommen, A. van der Laan, and H. M. Veenendaal-Hesselman, *J. Chromatogr.*, 1989, **481**, 452.
[180] W. J. Blanchflower and D. G. Kennedy, *Analyst (London)*, 1989, **114**, 1013.
[181] M. Kazacos and V. Mok, *Aust. J. Dairy Technol.*, 1986, **41**, 82.

The nitrobenzamide drugs aklomide, dinitolmide, and nitromide, together with dinsed and ethopabate, were determined[182] in chicken liver by extraction into ethyl acetate. Residues were cleaned up by solvent partition and gel chromatography on Sephadex LH-20 in solvent systems containing benzene. After alumina column chromatography the drugs were separated by RP-HPLC with an acetonitrile–water gradient and detected by UV at 260 nm. The nitroamides were spiked into frozen liver because metabolism was rapid at room temperature. Recoveries for the nitroamides were 90–100%, for ethopabate 70%, and for dinsed 50–60% at 200–1000 $\mu g\,kg^{-1}$. LoDs were about 50 $\mu g\,kg^{-1}$. Dinitolmide[183] and its metabolites 3-amino-2-methyl-5-nitrobenzamide and 5-amino-2-methyl-3-nitrobenzamide were extracted from chicken muscle and liver with dichloromethane–ethyl acetate $(1 + 1)$, which was cleaned up on neutral alumina. Analysis was via RP-HPLC eluted with pH 3.5 formate + EDTA buffer–methanol and reductive electochemical detection at −0.85 V against Ag/AgCl. The LoDs were 100 $\mu g\,kg^{-1}$.

2.11.2 Nitroimidazoles

Metabolism of these compounds is rapid. For example, radiolabelled dimetridazole (DMZ, Figure 15) administered to turkeys at 32 000 $\mu g\,kg^{-1}$ in water gave rise to less than 30 $\mu g\,kg^{-1}$ of residue after 2 days withdrawal.[184] The major metabolic pathway involves hydroxylation to produce 2-hydroxymethyl-1-methyl-5-nitroimidazole (HDMZ), followed by further oxidation or conjugation. In contrast to the nitrofurans, the nitro group is resistant to reduction and more than 90% of the excreted drug consists of various 5-nitroimidazoles.[184]

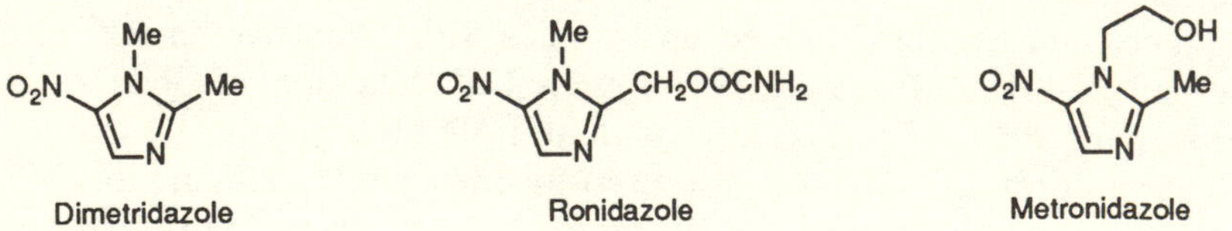

Dimetridazole   Ronidazole   Metronidazole

Figure 15

The drugs are moderately volatile and care should be exercised when evaporating solutions. Dilute working standards in water are relatively unstable and should be prepared weekly. Residues of DMZ were reported[185] to be stable for at least 3 months in pig muscle stored at −30 °C.

[182] K. D. Gallicano, H. Park, J. Yee, L. M. Young, and P. W. Saschenbrecker, *J. Assoc. Offic. Anal. Chem.*, 1988, **71**, 48.
[183] O. W. Parks and R. C. Doerr, *J. Assoc. Offic. Anal. Chem.*, 1986, **69**, 70.
[184] G. L. Law, G. P. Mansfield, D. F. Muggleton, and E. W. Parnell, *Nature* (*London*), 1963, **197**, 1024.
[185] G. Carignan, W. Skakum, and S. Sved, *J. Assoc. Offic. Anal. Chem.*, 1988, **71**, 1141.

Some multi-residue methods include nitroimidazoles but LoDs are often not adequate. An HPLC method for DMZ residues in poultry muscle and liver, with an LoD conservatively given as $10\ \mu g\,kg^{-1}$, was devised by Hobson-Frohock and Reader,[186] and successfully collaboratively tested.[187] Extensive partitioning of DMZ between acid and organic phases was carried out with pH adjustment. Determination was by RP-HPLC with a mobile phase at *ca.* pH 5 and UV detection at 315 nm. The extracts obtained are extremely clean and there is potential for improved sensitivity. The distribution coefficients of DMZ between acidic or basic aqueous solutions and organic phases are not favourable and in order to obtain an adequate recovery it is necessary to carry out extractions with up to five aliquots of solvent; this reduces sample throughput. Ronidazole is considerably more polar and recoveries with this procedure are low.

A more rapid and sensitive method for DMZ using reductive electrochemical detection has been reported by Carignan *et al.*,[185] and also applied to the determination of HDMZ.[172] Extraction into dichloromethane was followed by defatting and analysis by RP-HPLC with a pH 5 mobile phase and an EC setting of $-0.9\ V$ against Ag/AgCl. Recovery at $0.27\ \mu g\,kg^{-1}$ was 52%. Careful deoxygenation of the mobile phase and sample was required for maximum sensitivity. Ronidazole and ipronidazole could be determined by adjustment of the mobile phase organic modifier concentration. A similar rapid method for DMZ and HDMZ in pig tissues has been reported[188] capable of achieving LoDs of $0.5\ \mu g\,kg^{-1}$. Extraction into dichloromethane, evaporation, and defatting with hexane was followed by C_{18} SPE clean-up and gradient RP-HPLC method with UV detection at 315 nm.

Gas chromatography conditions for the analysis of DMZ have been established[189] and Newkirk *et al.*[190] analysed HDMZ in pig muscle by packed-column GC–ECD. The frozen tissue was extracted with ethyl acetate and basic phosphate buffer. Residues were purified by acid–base partition and silica SPE and then acetylated. Stainless steel frits were used in the SPE cartridge to eliminate an interference extracted from the standard polyethylene type. The LoD was below $1\ \mu g\,kg^{-1}$.

An optimized mobile phase of 5:95 acetonitrile–water (containing 20 mM sodium acetate at pH 5) for RP-HPLC of metronidazole on Hypersil C_{18} was determined by Zoest and co-workers.[191] Ipronidazole and

[186] A. Hobson-Frohock and J. A. Reader, *Analyst (London)*, 1983, **108**, 1091.

[187] Analytical Methods Committee, *Analyst (London)*, 1985, **110**, 1391.

[188] R. K. P. Patel, R. M. Parker, M. Chaplin, and H. Hassanali, in Proc. Euroresidue Conf. Residue Vet. Drugs Food., ed. N. Haagsma, A. Ruiter, and P. B. Czedik-Eysenberg, University of Utrecht, Utrecht, 1990, p. 294.

[189] L. Di Simone, F. Ponti, G. Settimi, and F. Martilloti, *Farmaco, Ed. Prat.*, 1981, **36**, 440.

[190] D. R. Newkirk, H. F. Righter, F. J. Schenck, J. L. Okrasinski, and C. J. Barnes, *J. Assoc. Offic. Anal. Chem.*, 1990, **73**, 702.

[191] A. R. Zoest, J. K. C. Lim, F. C. Lam, and C. T. Hung, *J. Liq. Chromatogr.*, 1988, **11**, 2241.

its hydroxlated metabolite were determined at 2 μg kg^{-1} in turkey tissue by MacDonald *et al.*, who found the drug to be pH and light sensitive.[192]

2.12   Quinolone Carboxylic Acids

Quinolone carboxylic acids (see Figure 16) are used extensively in fish farming. Nalidixic acid (NXA) at high concentration in serum has been monitored by gas chromatography after esterification with methyl iodide[193] but this approach seems not to have been applied to residue analysis. Most of the drugs are fluorescent and are analysed by HPLC to take advantage of the selective detection this makes possible. One multi-residue method[30] incorporates these drugs but with LoDs of 20 μg kg^{-1} for oxolinic acid (OXA) and 40 μg kg^{-1} for NXA.

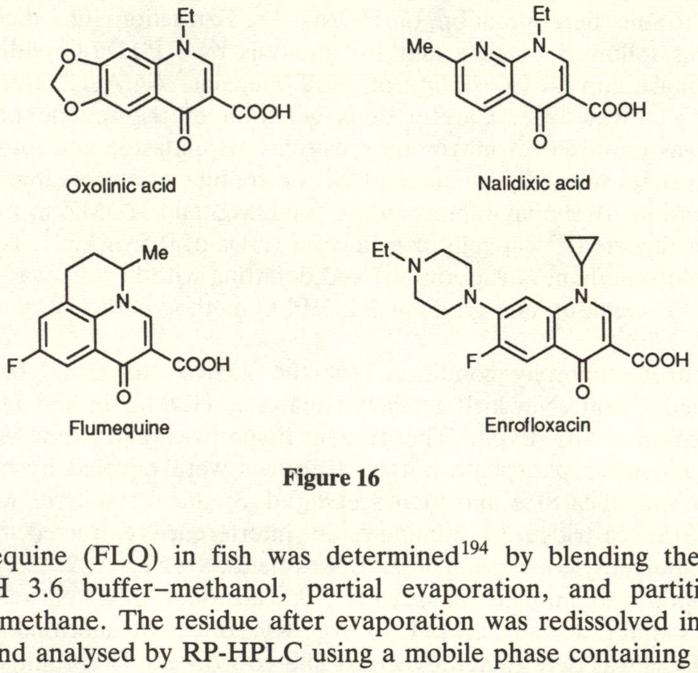

Oxolinic acid

Nalidixic acid

Flumequine

Enrofloxacin

Figure 16

Flumequine (FLQ) in fish was determined[194] by blending the muscle with pH 3.6 buffer–methanol, partial evaporation, and partition into dichloromethane. The residue after evaporation was redissolved in pH 9.0 buffer and analysed by RP-HPLC using a mobile phase containing aqueous citric acid. Fluorescence detection was employed. The LoD was 5 μg kg^{-1} and recovery at 25 μg kg^{-1} was 88%. A C$_6$ column exhibited peak tailing. Rogstad *et al.*[195] measured OXA and FLQ in fish muscle by blending with sodium hydroxide and acetone. The acidified supernatant was extracted into chloroform and again cycled between alkali and chloroform. SPE

[192] A. MacDonald, G. Chen, M. Kaykaty, and J. Fellig, *J. Agric. Food. Chem.*, 1971, **19**, 1222.

[193] R. H. A. Sorel and H. Roseboom, *J. Chromatogr.*, 1979, **162**, 461.

[194] O. B. Samuelsen, *J. Chromatogr.*, 1989, **497**, 355.

[195] A. Rogstad, V. Hormazabal, and M. Yndestad, *J. Liq. Chromatogr.*, 1989, **12**, 3073.

techniques were evaluated but found to be inferior, particularly with regard to sample throughput. The liquid–liquid partition method allowed a rate of 10–12 samples per day but only eight samples per day could be achieved with SPE, which did, however, include laboratory packing of empty cartridges. Liver was prepared in a similar manner but required cyano SPE. LoDs were 0.5 and $2 \mu g kg^{-1}$ for OXA and FLQ respectively. Recoveries were nearly quantitative at $10 \mu g kg^{-1}$.

Dialysis clean-up for OXA and FLQ was employed by Andresen and Rasmussen,[196] who extracted salmon liver with pH 10 buffer and hexane. The aqueous phase was injected into an automated dialyser and the recipient channel pulsed with pH 5 buffer to transport analytes into a polymer pre-concentration column. The analytical column, also polymer, was eluted with an acetonitrile–THF–phosphoric acid (20 mM) mobile phase. A fluorescence detector was used and the total HPLC cycle time was 14 minutes. The method was validated at $50 \mu g kg^{-1}$ and LoDs were estimated as $4 \mu g kg^{-1}$ for OXA and $7 \mu g kg^{-1}$ for FLQ. The method has also been applied to salmon muscle.[197] Direct analysis of muscle extracts on the polymer reverse phase column was acceptable at high analyte concentrations, but co-elution of drugs and co-extractives caused fluorescence quenching. These interferences were completely eliminated by dialysis. The system was readily reproduced in this laboratory and is very rapid. It is probably the method of choice for monitoring residues of these analytes, particularly if it could be extended to other quinolone carboxylic acids including the newer drugs such as enrofloxacin.

Sved *et al.* devised a rapid method for OXA in salmon muscle which involved blending tissue with sodium sulphate, extraction with ethyl acetate, and partition of the evaporated residue between hexane and 10 mM aqueous oxalic acid.[198] The aqueous phase was analysed directly by RP-HPLC in an oxalic acid-containing buffer with fluorescence detection. The LoD was $5 \mu g kg^{-1}$ and sample throughput in this laboratory was found to be about 30 per analyst per day.

OXA, nalidixic (NXA), and piromidic (PMA) acids were determined in eel[199] by methanol extraction, acid–base partition, and ion-pair RP-HPLC. Detection was by UV, which produced a claimed LoD of *ca.* $25 \mu g kg^{-1}$ for each analyte. The chromatograms presented showed significant peak tailing. Recoveries at $1000–2000 \mu g kg^{-1}$ were 80–90%. Another method for the same three analytes was given by Horie and colleagues,[200] who extracted fish with 0.1% metaphosphoric acid–methanol (3:2) followed by

[196] A. T. Andresen and K. E. Rasmussen, *J. Liq. Chromatogr.*, 1990, **13**, 4051.

[197] H. H. Thanh, A. T. Andresen, T. Agaster, and K. E. Rasmussen, *J. Chromatogr.*, 1990, **532**, 363.

[198] S. Sved, L. Larocque, A. Weninger, M. Schnurr, and D. Vas, in Proc. Euroresidue Conf. Residue Vet. Drugs Food., ed. N. Haagsma, A. Ruiter, and P. B. Czedik-Eysenberg, University of Utrecht, Utrecht, 1990, p. 356.

[199] S. Horii, C. Yasuoka, and M. Matsumoto, *J. Chromatogr.*, 1987, **388**, 459.

[200] M. Horie, K. Saito, Y. Hoshino, N. Nose, E. Mochizuki, and H. Nakazawa, *J. Chromatogr*,. 1987, **402**, 301.

C_{18} SPE clean-up. Analysis was by RP-HPLC with a mobile phase of 5 mM sodium dihydrogen phosphate–acetonitrile (3:2). Fluorescence detection was used except for PMA, where it was necessary to employ UV detection at 280 nm; the UV and fluorescence spectra are presented. Recoveries of 80% were achieved at 200 μg kg^{-1} and LoDs of 10 μg kg^{-1} reported.

Incorporation of 5 mM phosphate buffer was found necessary to obtain separation of the three analytes but variation of pH in the range 2.5–5.0 had no effect. High free silanol concentrations in the column packing were considered to lead to peak tailing and a wide-pore C_{18} column was found to give best performance. HPLC peaks considered to contain OXA were examined by MS, presumably using direct probe insertion. No details were given of the mass of OXA required. Further optimization of reverse phase chromatography was reported by Ikai *et al.*,[201] who developed a rapid (30 minute) method for OXA, NXA, and PMA. Fish was blended with anhydrous sodium sulphate and hexane–ethyl acetate (1:3) and the organic phase cleaned up on an amino SPE cartridge. RP-HPLC and UV detection were used with a mobile phase containing 10 mM oxalic acid to control peak tailing. Selection of an end-capped column was also considered important to minimize peak tailing. Recoveries at 100 μg kg^{-1} were 74–95%; the LoD of 50 μg kg^{-1} might have been improved using a fluorimeter.

OXA together with PMA and its 2- and 3-hydroxy metabolites were determined in fish[202] by blending with pH 6 phosphate buffer and ethyl acetate, evaporation of the organic layer, dissolution in acetonitrile, dilution (1:50) with 2% saline, and defatting with hexane, followed by extraction into chloroform and final partition into pH 10 borate buffer. Analysis was by ion-pair RP-HPLC, using pH 6.4 phosphate buffer–methanol (14:11) containing 0.05% trimethylammonium bromide, with UV detection at 330 nm. The two hydroxylated metabolites were shown to have spectra very similar to that of the parent compound.

An internal-surface reverse phase column was used for determination of OXA in fish tissue and plasma;[203] muscle and liver were extracted and cleaned up on a C_{18} SPE cartridge before analysis. Although the system permits direct injection of serum, the columns are expensive and offer no real advantage for deproteinized tissue residues.

A method for enrofloxacin in poultry has been presented[204] which requires soxhlet extraction of tissue for 15 hours with dichloromethane–methanol. After extensive acid–base partition, extracts were evaporated in the presence of diethylene glycol carrier to prevent losses. This clean-up allowed 82–93% recovery from chicken and turkey liver, muscle,

[201] Y. Ikai, H. Oka, N. Kawamura, M. Yamada, K.-I. Harada, M. Suzuki, and H. Nakazawa, *J. Chromatogr.*, 1989, **477**, 397.
[202] Y. Kasuga, A. Sugitani, and F. Yamada, *J. Food Hyg. Soc. Jpn.*, 1983, **24**, 484.
[203] H. V. Bjorklund, *J. Chromatogr.*, 1990, **530**, 75.
[204] T. B. Waggoner and M. C. Bowman, *J. Assoc. Offic. Anal. Chem.*, 1987, **70**, 813.

and skin with an LoD of 20 μg kg^{-1} as measured by direct spectrofluori-metric quantitation of the drug. RP-HPLC of the drug was carried out in a pH 3.5 mobile phase containing triethylamine with fluorescence detection at 282/445 nm. An LoD was not given but the traces from 50 μg kg^{-1} spikes were very clean and determination of 5 μg kg^{-1} was clearly possible.

Enrofloxacin and its metabolite ciprofloxacin in trout[205] were extracted with ethanol–acetic acid (99:1); the supernatant was made alkaline, de-fatted with hexane, and cleaned up on C$_{18}$ SPE for RP-HPLC analysis using UV detection at 278 nm. LoDs of 10 μg kg^{-1} were obtained. Incor-poration of acetic acid in the extractant was necessary to obtain good recovery of ciprofloxacin.

FLQ in pig muscle and kidney[206] was extracted with methanol containing 0.1% acetic acid, which was defatted with hexane, evaporated, and the residue separated by RP-HPLC and detected by fluorescence at 320/380 nm. The indicated LoD for kidney was 9 μg kg^{-1}. A C$_8$ column eluted with acetonitrile–phosphoric acid (25 mM) (7:3, adjusted to pH 3.5 with triethylamine) gave an excellent peak profile for standard FLQ, but residues were eluted relatively early amid co-extractives which may restrict quantitation at low concentrations. Recovery from kidney was poor, possibly due to losses at the defatting stage.

2.13 Sulphonamides and Diaminopyrimidines

Many sulphonamides are employed as veterinary drugs, often synergistic-ally with diaminopyrimidines. Selected structures are shown in Figure 17. They are widely used as feed additives, particularly for pigs, and with the more persistent sulphonamides, notably sulphamethazine (SMZ, also called sulphadimidine), there is a clear potential for residues to occur. The target tissue is kidney or liver. MRLs are set typically at 100 μg kg^{-1} for total sulphonamides and most current methods provide LoDs of 10–50 μg kg^{-1} for individual compounds.

Sulphonamides are subject to metabolism along various routes, (see Figure 18) including acetylation of the terminal N^4-amino group.[207] The acetyl compound may be hydrolysed to the free sulphonamide[208] by heating to 100 °C for 15 min in 1 M hydrochloric acid. Sheth and col-leagues[209] have demonstrated the reversible binding of sulphathiazole (STZ) via the N^4-amino group to reducing sugars as Amadori and

[205] B. Brinkmann, N. Haagsma, and H. Büning-Pfaue, in Proc. Euroresidue Conf. Residue Vet. Drugs Food, ed. N. Haagsma, A. Ruiter, and P. B. Czedik-Eysenberg, University of Utrecht, Utrecht, 1990, p. 123.
[206] L. Ellerbroek and M. Bruhn, *J. Chromatogr.*, 1989, **495**, 314.
[207] W. F. Rehm, K. Teelman, and E. Weidekamm, in 'Drug Residues in Animals', ed. A. G. Rico, Academic Press, New York, 1986, p. 65.
[208] A. J. Manuel and W. A. Steller, *J. Assoc. Offic. Anal. Chem.*, 1981, **64**, 794.
[209] H. B. Sheth, V. A. Yaylayan, N. H. Low, M. E. Stiles, and P. Sporns, *J. Agric. Food Chem.*, 1990, **38**, 1125.

Maillard compounds. Also, Parks[210] has presented evidence for the formation of the acid-unstable N^4-glucopyranosyl derivative in pig liver during frozen storage. Metabolism also proceeds by loss of the N^4-amino group. It has been reported[211] that storage of tissue at $-6\,^\circ$C led to a gradual decline in the concentration of detectable SMZ in the raw meat but after cooking the original level was restored. Parks has noted that many sulphonamide drugs are susceptible to photodegradation in the presence of riboflavin,[212] which will occur in crude extracts of liver. This may explain occasional anomalously low results but does not normally present a problem.

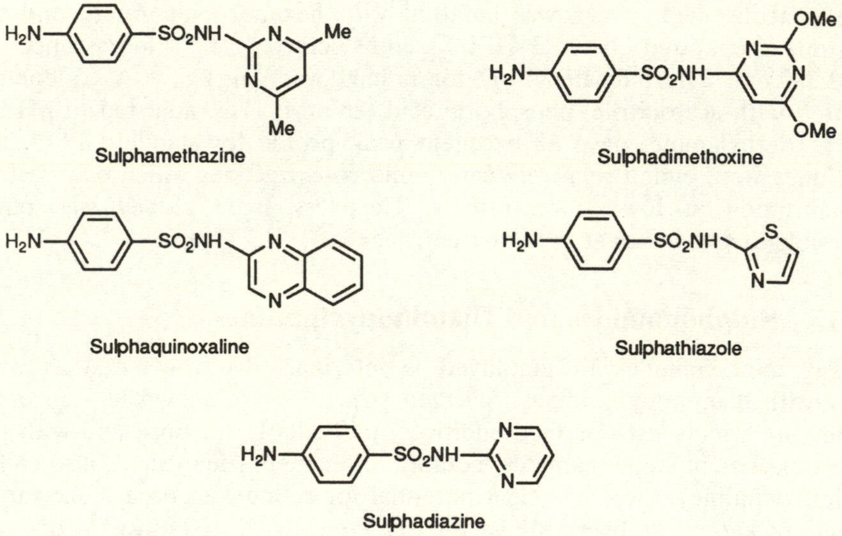

Sulphamethazine

Sulphadimethoxine

Sulphaquinoxaline

Sulphathiazole

Sulphadiazine

Figure 17

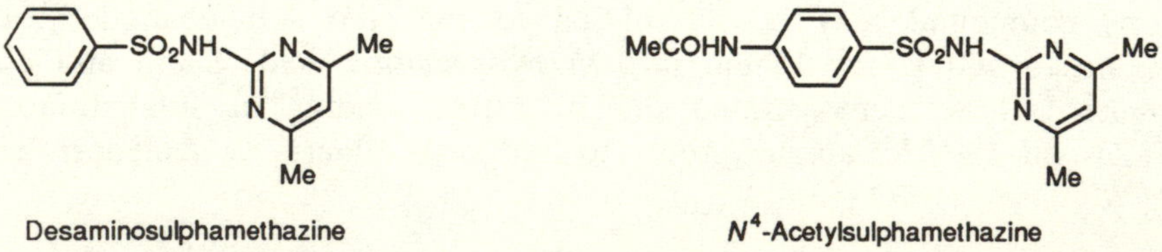

Desaminosulphamethazine

N^4-Acetylsulphamethazine

Figure 18

[210] O. W. Parks, *J. Assoc. Offic. Anal. Chem.* 1984, **67**, 566.
[211] T. Hassett, A. L. Patey, and G. Shearer, in Proc. Euroresidue Conf. Residue Vet. Drugs Food, ed. N. Haagsma, A. Ruiter, and P. B. Czedik-Eysenberg, University of Utrecht, Utrecht, 1990, p. 211.
[212] O. W. Parks, *J. Assoc. Offic. Anal. Chem.*, 1985, **68**, 1232.

Methodology for sulphonamides up to 1980 was reviewed by Horwitz.[213] This article will concentrate on later contributions. There is no shortage of sulphonamide procedures as the compounds are readily extracted from tissue with dichloromethane or similar solvents and chromatograph easily by TLC, HPLC, and, after derivatization, by GC. Sulphonamides have been successfully included in several multi-analyte RP-HPLC procedures (see Section 2.1). There are also MS methods which may be applied directly to crude sample extracts. The Charm receptor method[24] detects sulphonamides with good sensitivity.

Current published methods predominantly involve HPLC, but GC is often necessary for MS confirmation of positive results. Derivatization of sulphonamides for GC usually involves N^1-methylation of the sulphonamido bridge with diazomethane, a procedure liable to give rise to small amounts of other N-methylated products.[214] The reaction gave the least amount of side products when carried out in the dark and when excess diazomethane was evaporated without heating.[215] Sulphonamides are inherently electron capturing and no advantage was found in acetylating tissue extracts with fluorinated reagents.[208]

Conditions for GC–ECD analysis and GC–MS–MS confirmation of analyte identity at $100~\mu g\,kg^{-1}$ in pig liver have been reported by Matusik *et al.*[216] A more standard GC–EI-MS method for confirmation of six sulphonamides in egg, fish muscle, and animal tissue was devised by Takatsuki and Kikuchi.[215] An elaborate sample purification scheme was required, both before and after methylation. LoDs were $30–50~\mu g\,kg^{-1}$. These workers discuss problems with irreversible adsorption of sulphonamides to silica gel, but there are few other reports of similar effects. Five sulphonamides were determined simultaneously[217] in crude pig liver extracts by tandem MS at the relatively insensitive LoD of $100~\mu g\,kg^{-1}$. Samples were introduced on a moving belt but sample throughput was only 25 per week.

Sulphonamides chromatograph well by RP-HPLC and an excellent discussion[218] of the optimization of conditions for multi-sulphonamide analyses has been presented by Aerts *et al.* Few sulphonamides, except sulphanilamide and to a lesser extent sulphaquinoxaline (SQ), possess useful native fluorescence[219] but adequately sensitive detection is achieved by UV at about 280 nm. Haagsma and her colleagues have produced

[213] W. Horwitz, *J. Assoc. Offic. Anal. Chem.* 1981, **64**, 104.

[214] J. Gilbert, J. R. Startin, M. J. Shepherd, and J. C. Mitchell, *J. Chromatogr.*, 1986, **356**, 206.

[215] K. Takatsuki and T. Kikuchi, *J. Assoc. Offic. Anal. Chem.*, 1990, **73**, 886.

[216] J. E. Matusik, R. S. Sternal, C. J. Barnes, and J. A. Sphon, *J. Assoc. Offic. Anal. Chem.*, 1990, **73**, 529.

[217] E. M. H. Finlay, D. E. Games, J. R. Startin, and J. Gilbert, *Biomed. Environ. Mass Spectrom.*, 1986, **13**, 633.

[218] M. M. L. Aerts, W. M. J. Beek, and U. A. Th. Brinkman, *J. Chromatogr.*, 1988, **435**, 97.

[219] M. Petz, *J. Chromatogr.*, 1987, **423**, 217.

several useful methods for sulphonamide analysis, recommending ultrasonication as an effective means of promoting extraction from tissue. A procedure for SMZ in pig liver and muscle[220] involved initial extraction into dichloromethane, dilution with petroleum ether, and isolation on a silica solid phase extraction (SPE) cartridge, which was air dried, the residues being transferred to a reverse phase SPE cartridge for additional clean-up.

An extension of this method[221] for the determination of SMZ and its N^4-acetyl and desamino metabolites in the same matrices at an LoD of 10 $\mu g\,kg^{-1}$ indicated that the second SPE cartridge was not necessary. The simplified procedure has been in use within this laboratory for many hundreds of samples and has proved reliable.[222] One problem was encountered with a batch of dichloromethane which caused low recoveries and the solvent is now distilled before use. A recent report describes the same problem, which was also experienced with ether.[215] A similar TLC procedure for SMZ, sulphadiazine, sulphanilamide, SQ, and sulphadoxine was also described by Haagsma *et al.*,[223] using fluorescamine as visualizing spray, which had the disadvantage of rapid fading.

A multi-residue isocratic HPLC method for the same group of sulphonamides in pig kidney and muscle was reported[224] with LoDs of 50 $\mu g\,kg^{-1}$. The method involves clean-up by strong cation exchange SPE with pH adjustment by ammonia gas prior to elution of the sulphonamides. The authors suggest that the method may be extended to additional sulphonamide drugs. Another multi-sulphonamide screening method[225] (six drugs) based on TLC has been successfully collaboratively tested[226] with spiked and incurred residues at 50–200 $\mu g\,kg^{-1}$. The drugs are extracted into ethyl acetate and cleaned up by acid–base partition before TLC and visualization with fluorescamine.

An MSPD procedure[13] with LoDs of 50 $\mu g\,kg^{-1}$ or less was developed for the simultaneous assay of eight sulphonamides in pork muscle using diode array UV–HPLC. The method is rapid and merits further attention. A similar procedure has been devised for sulphadimethoxine (SDMO) at levels down to 50 $\mu g\,kg^{-1}$ in catfish.[227] Neither method has been validated with incurred tissue.

[220] N. Haagsma, R. J. Nooteboom, B. G. M. Gortemaker, and M. J. Maas, *Z. Lebensm.-Unters. Forsch.*, 1985, **181**, 194.

[221] N. Haagsma, H. G. J. M. Pluijmakers, M. M. L. Aerts, and M. J. Maas, *Biomed. Chromatogr.*, 1987, **2**, 41.

[222] W. W. F. Farrington and J. Tarbin, unpublished results.

[223] N. Haagsma, B. Dieleman, and B. G. M. Gortemaker, *Vet. Quart.*, 1984, **6**, 8.

[224] N. Haagsma and C. van de Water, *J. Chromatogr.*, 1985, **333**, 256.

[225] M. H. Thomas, K. E. Soroka, and S. H. Thomas, *J. Assoc. Offic. Anal. Chem.*, 1983, **66**, 881.

[226] M. H. Thomas, R. L. Epstein, R. B. Ashworth, and H. Marks, *J. Assoc. Offic. Anal. Chem.*, 1983, **66**, 884.

[227] A. R. Long, L. C. Hsieh, M. S. Malbrough, C. R. Short, and S. A. Barker, *J. Assoc. Offic. Anal. Chem.*, 1990, **73**, 868.

The most important matrix for sulphonamide determinations is currently pig tissue, because of the residue problems experienced, but other sample types have also been analysed. SQ in chicken muscle is determined in this laboratory in muscle at 20 $\mu g\,kg^{-1}$ by a modification of the method of Haagsma and van de Water,[224] replacing ammonia gas pH adjustment and methanol elution by treatment with 25% aqueous ammonia in methanol.[222] Eggs are extracted with ethyl acetate before clean-up and analysis in the same manner. The analysis of sulphanitran in chicken tissue[170] has been considered together with other nitro-drugs.

Weber and Smedley[228] reported a method for SMZ, SDMO, and SQ in milk with LoDs of about 5 $\mu g\,kg^{-1}$. The authors comment that cross-contamination can be a problem at these low levels and they routinely rinse glassware before use with 1 M hydrochloric acid, water, and methanol. In a subsequent publication they demonstrated the analysis of 10 sulphonamides in milk at 5–20 $\mu g\,kg^{-1}$, although a number of the drugs gave relatively low and variable recoveries.[229] Two isocratic separations were required. STZ has been measured in honey using TLC[230] and HPLC.[231]

There have been several literature reports of the development of immunoassays for SMZ.[232] Various manufacturers (see Table 4) have supplied immunological based reagents for sulphonamide analysis for some years; there are published data on the characteristics of the Idetek (Novo) assay applied to diluted pig plasma.[233] The LoD was 10 $\mu g\,kg^{-1}$ and the inter-well coefficient of variation at 100 $\mu g\,kg^{-1}$ was 6.5%. Intra-assay and inter-assay CVs of 10–18% and 9% respectively were reported for samples in the range 50–250 $\mu g\,kg^{-1}$. Cross-reactivities of 12% with sulphamerazine and <1% with other sulphonamides were reported.

Antibodies to SMZ are typically generated by conjugation of the parent drug through its N^4-amino function to carrier protein. Although antibodies are normally highly selective for SMZ, it is important to note that metabolites, such as the N^4-acetyl compound,[234] and degradation products modified in this region of the molecule will also be bound by the antibody. This may give rise to disagreement between immunoassay and chromatographic methods of analysis. With STZ it was found[209] that the Amadori adduct was up to 33 times more effective at binding antibody than the parent drug.

A number of procedures for the simultaneous determination of sulphonamides, diaminopyrimidines, and other drugs have been discussed in Section 2.1. Trimethoprim (TMP), Figure 19, is the most important

[228] J. D. Weber and M. D. Smedley, *J. Assoc. Offic. Anal. Chem.*, 1989, **72**, 445.
[229] M. D. Smedley and J. D. Weber, *J. Assoc. Offic. Anal. Chem.*, 1990, **73**, 875.
[230] E. Neidert, Z. Baraniak, and A. Suave, *J. Assoc. Offic. Anal. Chem.*, 1986, **69**, 641.
[231] C. P. Barry and G. M. MacEachern, *J. Assoc. Offic. Anal. Chem.*, 1983, **66**, 4.
[232] D. E. Dixon-Holland and S. E. Katz, *J. Assoc. Offic. Anal. Chem.*, 1988, **72**, 447.
[233] P. Singh, B. P. Ram, and N. Sharkov, *J. Agric. Food Chem.*, 1989, **37**, 109.
[234] J. R. Fleeker and L. J. Lovett, *J. Assoc. Offic. Anal. Chem.*, 1985, **68**, 172.

diaminopyrimidine in the UK, but others are used elsewhere. A 1972 procedure for pyrimethamine in chicken tissues[235] employing GC–ECD with an LoD of 100 $\mu g\,kg^{-1}$ appears to be the only chromatographic single-residue method for a diaminopyrimidine.

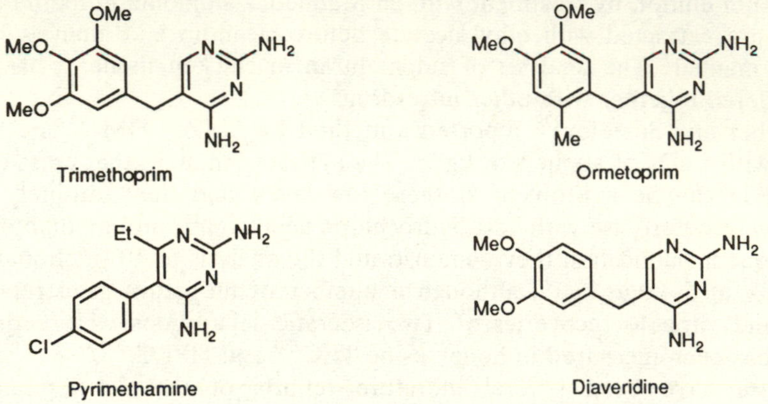

Trimethoprim

Ormetoprim

Pyrimethamine

Diaveridine

Figure 19

Diaveridine and SQ were determined together in chicken tissue and eggs by Sakano *et al.*[236] at levels down to 20 $\mu g\,kg^{-1}$. Ormetoprim and SDMO at equal LoDs of 200 $\mu g\,kg^{-1}$ were monitored in salmon[237] using acetonitrile extraction in the presence of TCA, followed by reverse phase SPE clean-up. Ormetoprim and SDMO were determined in cattle, chicken, and catfish tissues[238] homogenized in pH 10 buffer by ion-pair extraction into dichloromethane, which was injected directly into a normal phase HPLC system with chloroform–methanol eluent containing a trace of ammonia. UV detection at 288 nm permitted an LoD of about 10 $\mu g\,kg^{-1}$. SMZ could interfere with the determination of ormetoprim but was easily differentiated by wavelength ratioing.

Aerts *et al.*[218] describe an interesting method for rapid screening of TMP, sulphonamides, dapsone, and chloramphenicol in milk, eggs, and tissue. Samples are extracted into saline and cleaned up by a laboratory-constructed automated dialysis system with on-line HPLC analysis. Milk could be analysed directly with LoDs of about 20 $\mu g\,kg^{-1}$; other matrices were restricted to sulphonamide determination because a post-column derivatization procedure was required in order to improve selectivity.

[235] P. C. Cala, N. R. Trenner, R. P. Buhs, G. V. Downing, J. L. Smith, and W. J. A. VandenHeuvel, *J. Agric. Food Chem.*, 1972, **20**, 337.
[236] T. Sakano, S. Masuda, and T. Amano, *Chem. Pharm. Bull.*, 1981, **29**, 2290.
[237] J. A. Walisser, H. M. Burt, T. A. Valg, D. D. Kitts, and K. M. McErlane, *J. Chromatogr.*, 1990, **518**, 179.
[238] G. Weiss, P. D. Duke, and L. Gonzales, *J. Agric. Food Chem.*, 1987, **35**, 905.

Dialysis recoveries were only 5–10%, although reproducible, but a commercially available instrument provides greater than 50% recovery of a wide range of analytes.

SDMO and sulphamethoxazole, their N^4-acetyl metabolites, TMP, and its 3'- and 4'-hydroxy metabolites were determined[239] in pig muscle and lung tissue by extraction into buffer and on-line aqueous size exclusion chromatography–RP-HPLC. Detection was by UV, with sulphonamides monitored at 270 nm and TMP and metabolites at 240 nm. LoDs were $<25~\mu g\,kg^{-1}$.

2.14 Tetracyclines

Like sulphonamides, tetracyclines (Figure 20) have been used extensively as veterinary drugs for many years. The most important members of the class are chlortetracycline (CTC), oxytetracycline (OTC), and the parent drug, tetracycline (TC). The target organ for analysis is kidney, with residue levels about three times higher than in liver.[240]

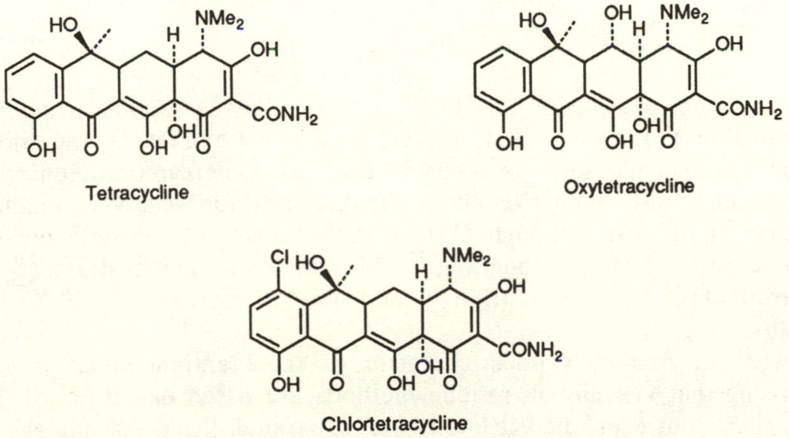

Tetracycline

Oxytetracycline

Chlortetracycline

Figure 20

Tetracyclines are stable indefinitely as the solid drugs. Standards in methanol may be stored for prolonged periods at low temperatures, but aqueous solutions are subject to decomposition.[241] Half-lives of OTC at room temperature range from a few days to a few hours as the pH increases from 2 to 10. Anhydrotetracyclines are produced on heating in

[239] M. J. B. Mengelers, J. F. Staal, M. M. L. Aerts, H. A. Kuiper, and A. S. J. P. A. M. van Miert, in Proc. Euroresidue Conf. Residue Vet. Drugs Food, ed. N. Haagsma, A. Ruiter, and P. B. Czedik-Eysenberg, University of Utrecht, Utrecht, 1990, p. 267.
[240] G. O. Korsrud and J. D. MacNeil, *Food Add. Contam.*, 1988, **5**, 149.
[241] P. Sporns, S. Kwan, and L. A. Roth, *J. Food Protect.*, 1986, **49**, 383.

1 M acid and, although anhydro-TC and -CTC are stable, anhydro-OTC reacts further, producing a mixture of α- and β-apo-OTC.[242] The structures of some of these products are shown in Figure 21.

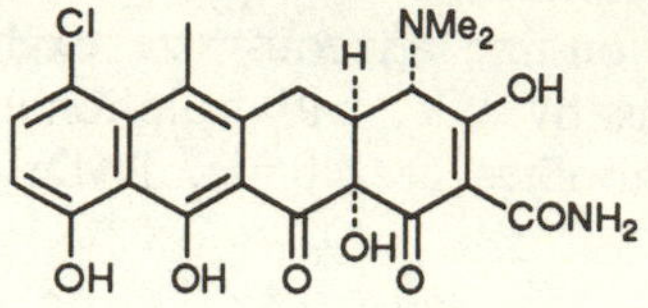

Anhydrochlortetracycline 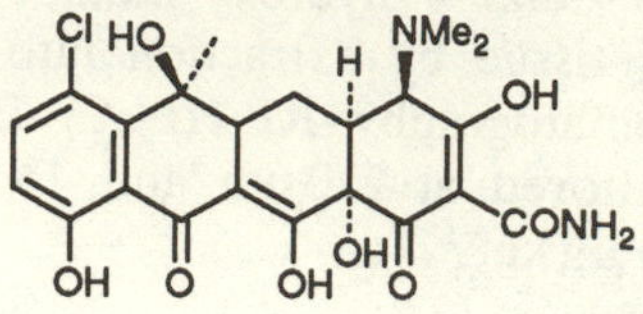   Epichlortetracycline

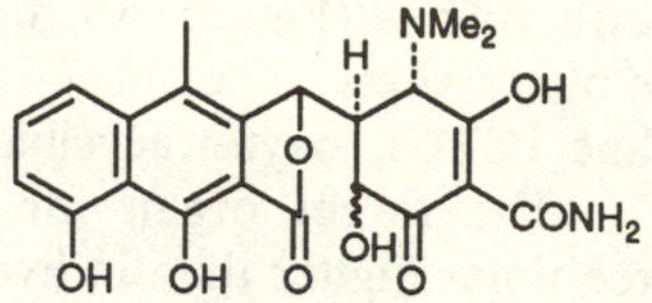

α- and β-Apooxytetracycline

Figure 21

It has been suggested[243] that OTC is particularly susceptible to photodecomposition and some workers protect final tissue extracts by the addition of an antioxidant, such as ascorbic acid or β-mercaptopropionic acid. Tetracyclines also epimerize about the C-4 position relatively readily in solution at pH 2–6 although OTC is rather resistant, possibly owing to intramolecular hydrogen bonding.[244] Murray *et al.* demonstrated[245] that incurred OTC residues in trout were stable to storage at −20 °C for 3 months.

Given the extremely polar character of these antimicrobials it is not surprising that virtually all residue methods are based on HPLC or TLC. They can, however, be silylated and separated by GC, and the MS fragmentation pattern of persilylated TC has been elucidated.[246] TLC detection may be carried out by UV densitometry or a number of spray reagents, the simplest and most effective of which appears to be 0.2 M magnesium chloride.[247] This produces fluorescent spots permitting visualization of less than 50 ng.

HPLC conditions for TC have been discussed by Hoogenmartens *et al.*[248] Excessive peak tailing, possibly due to complexation of the drugs

[242] R. B. Ashworth, *J. Assoc. Offic. Anal. Chem.*, 1985, **68**, 1013.
[243] B. Scales and D. A. Assinder, *J. Pharm. Sci.*, 1973, **62**, 913.
[244] N. H. Khan, E. Roets, J. Hoogmartens, and H. Vanderhaeghe, *J. Chromatogr.*, 1987, **405**, 229.
[245] J. Murray, A. S. McGill, and R. Hardy, *Food Add. Contam.*, 1988, **5**, 77.
[246] K. Tsuji and J. H. Robertson, *Anal. Chem.*, 1973, **45**, 2136.
[247] A. Szabo, M. K. Nagy, and E. Tomorkeny, *J. Chromatogr.*, 1978, **151**, 256.
[248] J. Hoogenmartens, E. Roets, and H. Vanderhaeghe, *Symp. Biol. Hung.*, 1988, **37**, 225.

with metallic impurities in packing materials, has in the past been a serious problem for tetracycline chromatography. Tailing is less of a concern with modern base-deactivated silicas, but packings from different manufacturers vary widely in their metal content and selection of a reverse phase material with a low iron concentration would be prudent. Peak profiles may be improved by addition to the mobile phase of complexing agents such as EDTA or other modifiers, particularly oxalic acid[249] which gives excellent results and can be recommended. It functions as a suppressor of silanol ionization and possibly also as a chelating agent. Alternatively, addition of ion-pairing compounds to the eluant may be helpful.[250]

It has sometimes been suggested that C_8 reverse phase columns provide better performnce than C_{18} types, perhaps because in the former the unmodified silanols are more accessible and participate in the separation mechanism, but both have been used successfully. Polymeric reverse phase packings have found favour recently.[244,250] It might also be useful to replace stainless steel filters and column-end frits with titanium versions, which are readily available at reasonable cost. Frits contribute by far the largest proportion of the total metal surface area in standard HPLC systems. Completely metal-free instruments could be used but this approach does not seem necessary.

Tetracyclines have UV absorbance maxima in the region of 360 nm and relatively high extinction coefficients. Thus UV detection is almost universally used. Alternatives include direct probe MS–MS determination on crude extracts,[251] with a claimed LoD of $1 \mu g \, kg^{-1}$. CTC has been quantitated by alkaline rearrangement to the fluorescent iso-CTC, which afforded[252] an LoD of $50 \mu g \, kg^{-1}$ in pig tissue.

Acidic extraction of samples is necessary. Moats[250] used 1 M hydrochloric acid and obtained recoveries of 90% from a range of cattle and pig tissues at $1000 \mu g \, kg^{-1}$, although Martinez and Shimoda[253] considered that losses of about 40% during feed analysis arose from conversion of CTC into its epimer due to the use of pH 2 buffer. Extract-acid contact conditions are likely to be critical. Other workers have successfully employed citrate–phosphate or succinate buffers at pH 4, almost invariably incorporating EDTA to prevent losses of tetracyclines by complex formation.

OTC and TC were determined in milk[254] at $10 \mu g \, kg^{-1}$ by acidification and precipitation of proteins with acetonitrile. After further extensive liquid–liquid partitioning the drugs were analysed by RP-HPLC. Columns were conditioned with CTC before use to eliminate tailing. A simple

[249] Y. Ikai, H. Oka, N. Kawamura, M. Yamada, K.-I. Harada, and M. Suzuki, *J. Chromatogr.*, 1987, **411**, 313.
[250] W. A. Moats, *J. Chromatogr.*, 1986, **358**, 253.
[251] P. Traldi, S. Daolio, B. Pelli, R. Maffei Facino, and M. Carini, *Biomed. Mass Spectrom.*, 1985, **12**, 493.
[252] W. J. Blanchflower, R. J. McCracken, and D. A. Rice, *Analyst (London)*, 1989, **114**, 421.
[253] E. E. Martinez and W. Shimoda, *J. Assoc. Offic. Anal. Chem.*, 1989, **72**, 848.
[254] D. J. Fletouris, J. E. Psomas, and N. A. Botsoglou, *J. Agric. Food Chem.*, 1990, **38**, 1913.

ultrafiltration of buffered milk was adequate for measurement of tetracyclines by RP-HPLC at $10–20\,\mu g\,kg^{-1}$ using a mobile phase containing oxalic acid.[255] Tetracyclines were determined in trout[256] with an LoD estimated at $10\,\mu g\,kg^{-1}$. Horie and colleagues devised a multi-residue[31] method also incorporating chloramphenicol, sulphonamides, and FZD, with claimed OTC and CTC LoDs of $50\,\mu g\,kg^{-1}$.

Sample clean-up is typically performed by C_{18} SPE. Extracts are loaded, washed with water, and eluted, frequently by methanol, although a Japanese group[249,257] find that C_{18} cartridges, air dried and washed with a small volume of ethyl acetate before elution with 5:95 or 10:90 methanol–ethyl acetate, provide cleaner extracts. They claim LoDs of $20–50\,\mu g\,kg^{-1}$ for seven tetracyclines in honey[257] and $10\,\mu g\,kg^{-1}$ for four tetracyclines in cattle and pig tissue.[249] Long *et al.* have demonstrated the MSPD technique for tetracyclines in milk[258] at levels down to $100\,\mu g\,kg^{-1}$ although problems were experienced with low and variable recoveries of TC. A similar procedure was devised for OTC in catfish muscle[259] and spiked samples were satisfactory at $50\,\mu g\,kg^{-1}$.

A novel method for tetracyclines was developed in this laboratory by Farrington *et al.* who turned the normally inconvenient feature of metal chelation to advantage by employing chelating Sepharose in Cu^{II} form as part of the clean-up.[260] Analysis was by RP-HPLC in 10 mM oxalic acid–acetonitrile and UV detection at 350 nm. Recoveries of 60–80% were obtained from pig and cattle kidney and muscle, milk, sheep kidney, and trout muscle at $50\,\mu g\,kg^{-1}$. LoDs are in the region of $10\,\mu g\,kg^{-1}$. The procedure has been used for over 500 samples. A loss of recovery, experienced when analysing extracts on HPLC instruments previously used with eluents containing acidic phosphate buffer, was overcome by passivizing the system with 50% aqueous nitric acid.

Levels of tetracyclines and other antibiotics have been analysed in North American abattoirs for many years using the microbiological Swab Test on Premises[261] (STOP). Confirmation has been carried out by a number of methods, including TLC–bioautography.[262] No immunological methods have yet appeared for this class of drugs, and the Charm receptor assay is relatively insensitive to them.

[255] M. H. Thomas, *J. Assoc. Offic. Anal. Chem.*, 1989, **72**, 564.

[256] I. Nordlander, H. Johnsson, and B. Osterdahl, *Food Add. Contam.*, 1987, **4**, 291.

[257] H. Oka, Y. Ikai, N. Kawamura, K. Uno, and M. Yamada, *J. Chromatogr.*, 1987, **400**, 253.

[258] A. R. Long, L. C. Hsieh, M. S. Malbrough, C. R. Short, and S. A. Barker, *J. Assoc. Offic. Anal. Chem.*, 1990, **73**, 379.

[259] A. R. Long, L. C. Hsieh, M. S. Malbrough, C. R. Short, and S. A. Barker, *J. Assoc. Offic. Anal. Chem.*, 1990, **73**, 864.

[260] W. H. H. Farrington, J. Tarbin, and J. Bygrave, *Food Add. Contam.*, 1991, **8**, 55.

[261] R. W. Johnston, R. H. Reamer, E. W. Harris, H. G. Fugate, and B. Schwab, *J. Food Protect.*, 1981, **44**, 828.

[262] E. Neidert, P. W. Saschenbrecker, and F. Tittiger, *J. Assoc. Offic. Anal. Chem.*, 1987, **70**, 197.

2.15 Thyrostats

Few chromatographic methods have been reported for thyrostats. Thiouracil and methylthiouracil (MTU) (Figure 22) were extracted from meat (presumably cattle muscle) with ethyl acetate and cleaned up using silica SPE before separation by RP-HPLC and UV or oxidative electrochemical detection.[263] LoDs of 10 μg kg^{-1} (EC) or 25 μg kg^{-1} (UV) were claimed; recoveries in the range 100–2000 μg kg^{-1} were 70%. A palladium–calcein fluorimetric spray reagent was used[264] to visualize four substituted thiouracils and 2-mercapto-1-methylimidazole in extracts of cattle thyroids after TLC. The LoD for MTU was stated to be 200 μg kg^{-1}.

Thiouracil

Methylthiouracil

Figure 22

2.16 Tranquilizers and Sedatives

The tranquilizers (Figure 23) are a rather varied group of basic compounds. Azaperol, a metabolite of azaperone, is also of interest because it may be oxidized back to the parent drug. Kidney is the target organ for carazolol and is acceptable for the other drugs, although for the promazine group liver is preferred. No losses of seven drugs were experienced from spiked kidney stored at −20 °C for one month.[265] Some of the tranquilizers are light-sensitive. Because of the number of compounds potentially used, multi-residue procedures are desirable for surveillance purposes.

Keukens and Aerts[265] developed a multi-residue method for acepromazine, azaperone (and its metabolite azaperol), carazolol, chlorpromazine, haloperidol (not often used as an animal drug and an internal standard candidate), propiopromazine, and xylazine in pig kidney. Tissue is homogenized and extracted with acetonitrile, which is subjected to clean-up by C$_{18}$ SPE. RP-HPLC analysis was carried out and had to be carefully optimized to avoid analyte adsorption and achieve resolution of critical pairs, including carazolol and azaperol. These two drugs were determined by fluorescence detection at 246/351 nm, and all drugs by UV at 240 nm. Xylazine has poor UV characteristics. Recoveries at the 10–40 μg kg^{-1}

[263] H. Hooijerink and W. G. De Ruig, *J. Chromatogr.*, 1987, **394**, 403.
[264] G. Moretti, M. Amici, P. Cammarata, and F. Fracassi, *J. Chromatogr.*, 1988, **442**, 459.
[265] H. J. Keukens and M. M. L. Aerts, *J. Chromatogr.*, 1989, **464**, 149.

level were 90–100% with the exception of xylazine (52%) and conservative LoDs were $0.3–6 \, \mu g \, kg^{-1}$. Sample throughput was 30 per day per analyst. Much useful information on the clean-up procedure is presented. No problems were experienced with adsorption of promazine drugs to glassware but vials were carefully conditioned. The method was used without problems for routine monitoring of over 1000 samples. It has significant advantages over other multi-tranquilizer methods reported.

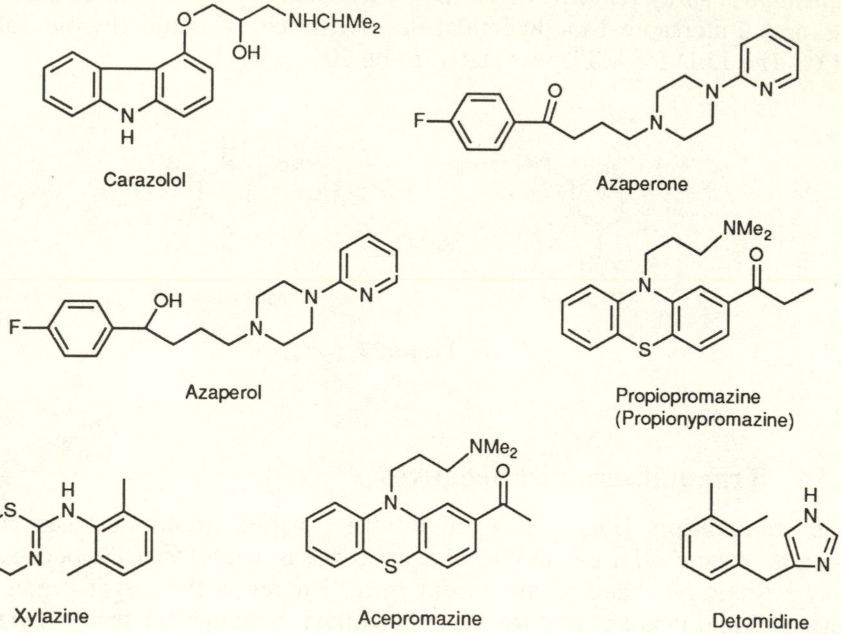

Carazolol

Azaperone

Azaperol

Propiopromazine
(Propionypromazine)

Xylazine

Acepromazine

Detomidine

Figure 23

Early LC and GC single-residue protocols for tranquilizers have been summarized by van Ginkel *et al.*,[266] who presented a method for the determination in pig kidney of the same seven tranquilizers listed above. Tissue was homogenized, incubated with sodium hydroxide, and then extracted with ether. The organic phase was cleaned up on a diol SPE cartridge. After a preparative RP-HPLC separation with UV detection at 235 nm, three fractions were further cleaned up by liquid–liquid partition and individual tranquilizers determined in various TLC systems. LoDs were $<1–10 \, \mu g \, kg^{-1}$. Partial confirmation of identity could be achieved by diode array spectrophotometry at concentrations two- or three-fold higher. Recoveries near the LoDs were about 60% except for xylazine where only 31% was achieved because of incomplete elution during SPE clean-up. The

[266] L. A. van Ginkel, P. L. W. J. Schwillens, and M. Olling, *Anal. Chim. Acta.*, 1989, **225**, 137.

method was reported to have been applied successfully to residue surveillance, where good agreement was recorded with results obtained by the procedure of Keukens and Aerts.[265]

Vogelgesang[267] has determined 10 tranquilizers in cattle and pig tissue. Samples were ultrasonicated with acid aqueous acetonitrile, the supernatant made alkaline after defatting with hexane, and analytes partitioned into dichloromethane. Carazolol was determined by gradient cyano RP-HPLC in acetonitrile–phosphoric acid with fluorescence detection. After additional column clean-up on deactivated basic alumina, tranquilizers were determined by GC–NPD although many other peaks appear in the chromatogram, particularly from liver extracts. Xylazine was particularly affected by co-extractives. The quality of acetonitrile used for extraction was important; HPLC grades contained interferences. Careful control of the alumina water content was critical. LoDs were about 10 μg kg^{-1} except for carazolol (0.04 μg kg^{-1}), azaperol (30 μg kg^{-1}), and xylazine (200 μg kg^{-1}).

Detomidine in horse muscle was blended with aqueous acid, the mixture adjusted to pH 12, and the drug extracted into hexane by pentafluorobenzoylation.[268] Capillary GC–negative ion CI-MS afforded an LoD of 0.2 μg kg^{-1}.

Other procedures for tranquilizers are available, but for fewer compounds, and they appear to offer no advantages over the multi-residue methods described above.

2.17 Other Drugs

This section covers a number of drugs which do not fall into the classes discussed above. The structures of some of these compounds are shown in Figure 24.

2.17.1 *Amprolium*

The target tissue is egg yolk. Amprolium has been determined in chicken tissue and eggs with low LoDs. Malisch[269] blended samples with TCA–methanol, defatted with t-butyl methyl ether, and employed ion-pair extraction followed by ion-pair RP-HPLC with UV detection. An LoD of 10 μg kg^{-1} was claimed, and confirmation by capillary GC–NPD and GC–MS of a sulphite cleavage product, 2-picolone, was demonstrated.

A method[270] for egg yolk or muscle employing dialysis clean-up after aqueous dilution or extraction, ion-pair RP-HPLC with a mobile phase containing triethylamine, and post-column oxidation to amprochrome with fluorimetric detection afforded an LoD of 3 μg kg^{-1}. Some membrane

[267] J. Vogelgesang, *Dtsch. Lebensm.-Rundsch.*, 1989, **85**, 251.
[268] L. Vuorilehto, J. S. Salonen, and M. Anttila, *J. Chromatogr.*, 1990, **530**, 137.
[269] R. Malisch, *Dtsch. Lebensm.-Rundsch.*, 1988, **84**, 282.
[270] W. van Leeuwen and H. W. van Gend, *Z. Lebensm.-Unters. Forsch.*, 1988, **186**, 500.

Amprolium

Clopidol

Ethopabate

Halofuginone

Levamisole
[L(−)-Tetramisole]

Novobiocin

Pyrantel

Morantel

Dapsone

Malachite Green

Rafoxanide

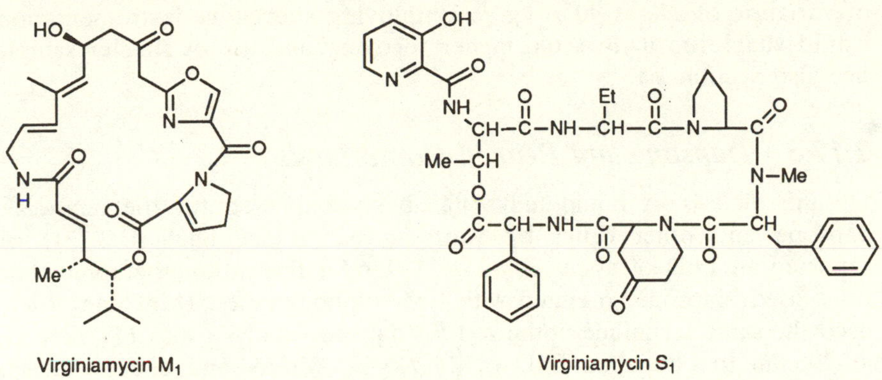

Virginiamycin M$_1$ Virginiamycin S$_1$

Figure 24

fouling was experienced and dialysis recovery was only 20%, but this could be improved.[11] The post-column system had been applied by Nagata and Saeki[271] after extraction of chicken muscle with methanol, defatting with hexane, and clean-up on basic alumina. An LoD of 10 μg kg^{-1} was reported.

2.17.2 Clopidol

The target organ is liver. Clopidol in chicken tissue was determined[272] by extraction with methanol, clean-up on basic alumina and anion exchange resin, esterification with propionic anhydride, and GC–ECD. The LoD was 0.1 μg kg^{-1}. Recovery at 0.5 μg kg^{-1} was 65%. Selection of anion exchanger was discussed. A similar method[273] employing RP-HPLC with UV detection at 270 nm afforded an LoD of 10 μg kg^{-1}.

Whole homogenized egg, white, or yolk were separately diluted, centrifuged, mixed on-line with aqueous sodium hydroxide, and dialysed against water in a continuous flow system.[274] The dialysate was passed through a concentration column and clopidol residues chromatographed on a C$_8$ RP-HPLC column, using methanol–pH 7.0 phosphate buffer (15 + 85). Retention of clopidol in 100% water was irreproducible unless the concentration column was constantly maintained under pressure. Dialysis recovery was 17%. UV detection at 270 nm enabled LoDs of 10–20 μg kg^{-1}, limited by interfering co-extractives rather than the dialysis efficiency. Reproducibility was excellent, as demonstrated by a coefficient

[271] T. Nagata and M. Saeki, *J. Assoc. Offic. Anal. Chem.*, 1986, **69**, 941.
[272] L.-G. Ekstrom and L. Kuivinen, *J. Assoc. Offic. Anal. Chem.*, 1984, **67**, 955.
[273] C. A. Mtema, H. Nakazawa, and E. Takabatake, *J. Assoc. Offic. Anal. Chem.*, 1984, **67**, 334.
[274] E. M. Mattern, C. A. Kan, and H. W. van Gend, *Z. Lebensm.-Unters. Forsch.*, 1990, **190**, 25.

of variation of 2% at 40 μg kg^{-1}. Employing alternative instrumentation[11] would enable up to five-fold higher recovery and use of smaller samples. See also Section 2.1.

2.17.3   Dapsone and Related Compounds

On-line dialysis of liquid milks has been employed for the analysis of dapsone and other drugs by Aerts *et al.*[218] (see Section 2.13), who reported an LoD of about 20 μg kg^{-1}. Under the chromatographic conditions used, dapsone co-eluted with two sulphonamides. Brinkman *et al.*[275] used the same technique optimized for dapsone analysis with UV detection at 296 nm to obtain an LoD of 2 μg kg^{-1}. Mono- and di-acetyldapsone metabolites were also determined with LoDs of 2 and 5 μg kg^{-1}. Full fat milk was defatted by freezing and milk powders were reconstituted as 10% aqueous solutions. The method was applied to a survey of 3500 samples without problem.

Sulphamoyldapsone in pig tissues[276] was extracted with hexane-saturated acetonitrile, defatted with hexane, and cleaned up with basic alumina. Care was required to avoid complete evaporation of residues during clean-up. Analysis by RP-HPLC in acetonitrile–methanol–water and UV detection at 292 nm afforded an LoD of 20 μg kg^{-1}. Bithionol sulphoxide and its oxidized and reduced metabolites in milk[277] were analysed by acidification and extraction into THF–ether. The residue was defatted and transferred to aqueous alkali before chromatography on a C$_8$ column in acetonitrile–0.2 M sulphuric acid. UV detection was at 303 nm and provided LoDs of 25 μg kg^{-1}.

2.17.4   Dyes

Malachite Green in rainbow trout[278] has been determined with an LoD of 1 μg kg^{-1}. Tissue was extracted using acetonitrile with perchloric acid deproteinization, defatted with hexane, and leuco-dye in the residue was oxidized with lead dioxide. RP-HPLC was in DMF–water–acetonitrile–methanol containing perchloric acid and naphthalene-1-sulphonic acid, with UV detection at 618 nm. The leuco-dye was the predominant residue.

Munns *et al.*[279] monitored Leucogentian Violet in chicken fat by extraction into dichloromethane and acid–base partition, followed by cyano RP-HPLC in pH 4.5 acetate/EDTA buffer–acetonitrile and EC

[275] M. B. C. Brinkman, H. W. van Gend, and E. M. Mattern, *Z. Lebensm.-Unters. Forsch.*, 1986, **183**, 97.

[276] Y. S. Endoh, R. Yamaoka, and N. Sasaki, *J. Assoc. Offic. Anal. Chem.*, 1987, **70**, 1031.

[277] D. Mourot, M. Dagorn, and B. Delepine, *J. Assoc. Offic. Anal. Chem.*, 1987, **70**, 810.

[278] K. Bauer, H. Dangschat, H. O. Knoeppler, and J. Neudegger, *Arch. Lebensmittelhyg.*, 1988, **39**, 97.

[279] R. K. Munns, J. E. Roybal, J. A. Hurlbut, and W. Shimoda, *J. Assoc. Offic. Anal. Chem.*, 1990, **73**, 705.

detection at 1.0 V against Ag/AgCl. Strong acid or base and metal contamination promote oxidation of residues to the coloured form, and irreversible adsorption on to glass surfaces occurred when solutions were evaporated to complete dryness.

2.17.5 Ethopabate

The drug was determined with an LoD of about 50 $\mu g\,kg^{-1}$ together with a group of nitrobenzamides[182] and at similar LoDs forms part of many multi-residue methods (see Section 2.1). Nagata and colleagues[280] list previous LC and GC methods for the drug and measured ethopabate in chicken tissues by extraction into acetonitrile and transfer to ethyl acetate, clean-up by Florisil column chromatography, and RP-HPLC in acetonitrile–pH 4.0 phosphate buffer containing triethanolamine with fluorescence detection at 306/350 nm. The fluorescence spectra are presented. Recovery at 10 $\mu g\,kg^{-1}$ was 88% with an LoD of 0.5 $\mu g\,kg^{-1}$.

2.17.6 Halofuginone

The target tissue is liver and residues are stable at $-18\,°C$ for at least one year.[281] Alkaline trypsinized chicken muscle, skin, fat, liver, or kidney was extracted with ethyl acetate[282] and partitioned into aqueous ammonium acetate which was cleaned up by C_{18} SPE. RP-HPLC was carried out in acetonitrile–acetate (pH 4.3) buffer, monitored by UV at 243 nm. The LoD was estimated at 1 $\mu g\,kg^{-1}$ and recoveries were 70–100% at 15–1030 $\mu g\,kg^{-1}$.

A collaborative study[281] of this method indicated that it was important to minimize the time for which the drug was in contact with ethyl acetate and to keep the organic phase cold. It was also necessary to evaporate dissolved ethyl acetate from the ammonium acetate buffer prior to SPE. Recoveries from muscle of 70–95% at 100 $\mu g\,kg^{-1}$ were reported by five laboratories. The LoD was considered to be 10 $\mu g\,kg^{-1}$.

2.17.7 Levamisole

The drug has been determined by GC–NPD, but it is very polar and peak tailing can be a problem. Careful inactivation of columns and elimination of metal surfaces is required to obtain relatively symmetrical peaks. Tissue or fat was blended with sodium hydroxide and extracted into heptane–isoamyl alcohol. After acid–base partition, residues were chromatographed

[280] T. Nagata, M. Saeki, H. Nakazawa, M. Fujita, and E. Takabatake, *J. Assoc. Offic. Anal. Chem.*, 1985, **68**, 27.

[281] Analytical Methods Committee, *Analyst (London)*, 1984, **109**, 171.

[282] A. Anderson, E. Goodall, G. W. Bliss, and R. N. Woodhouse, *J. Chromatogr.*, 1981, **212**, 347.

on a packed OV-17 column. An LoD of 5 μg kg^{-1} was claimed[283] and recovery at 1000 μg kg^{-1} was 86%. A similar clean-up was used by Stout *et al.*[284] for GC–EI-MS confirmation of levamisole in cattle and pig liver at levels down to 50 μg kg^{-1}. A number of interferences were found to originate from soap residues on glassware and by migration from plastic materials. Milk was analysed[285] by adjustment to pH 4.6, addition of methanol, and heating. The supernatant was applied to an Extrelut column, defatted, and then eluted with dichloromethane. Determination was by RP-HPLC in methanol–50 mM ammonium carbonate (65:35) with UV detection at 220 nm. Recovery at 200 μg kg^{-1} was 76% and the LoD was 50 μg kg^{-1}.

2.17.8 Morantel and Pyrantel

Current methods for these two closely related drugs involve hydrolysis of their metabolites and the determination of degradation products.[286] Each gives rise to two measurable compounds. One, *N*-methylpropanediamine (MAPA), is produced from both; pyrantel in addition forms 3-(2-thienyl)-acrylic acid (TAA), while 3-(3-methyl-2-thienyl)acrylic acid (MTAA) originates from morantel. Factors for conversion of MAPA and TAA into drug equivalents have been described.[287] Standard solutions of the parent drugs are stable if stored with exclusion of light. Liver is the target tissue.

Milk[288] was analysed by hydrolysis overnight at 110 °C with sodium hydroxide to produce MAPA, which was extracted into toluene and back-extracted into acid. This solution was buffered to pH 9.4 and derivatized, also overnight, with 4-fluoro-3-nitrotrifluoromethylbenzene. The product was purified by TLC and quantitated by packed column GC–ECD. A morantel analogue producing *N*-ethylpropanediamine was used as an internal standard, and absolute recovery and LoD information were not given. A coefficient of variation of 12% and a relative recovery of 88% at 12.5 μg kg^{-1} were reported. Blank milk gave a morantel equivalent of 3 μg kg^{-1}, implying an LoD of *ca.* 10 μg kg^{-1}.

For determination of the substituted acrylic acids, acid-digested milk was made alkaline and extracted into toluene. After transfer into aqueous acid, residues were hydrolysed with sodium hydroxide to produce MTAA and TAA.[286] Following acid–base partition, the acids was determined by RP-HPLC with UV detection at 313 nm. MTAA and TAA identity could

[283] R. Woestenborghs, L. Michielsen, and J. Heykants, *J. Chromatogr.*, 1981, **224**, 25.
[284] S. J. Stout, A. R. DaCunha, R. E. Tondreau, and J. E. Boyd, *J. Assoc. Offic. Anal. Chem.*, 1988, **71**, 1150.
[285] B.-G. Osterdahl, H. Johnsson, and I. Nordlander, *J. Chromatogr.*, 1985, **337**, 151.
[286] M. J. Lynch, F. R. Mosher, L. A. Brunner, and S. R. Bartolucci, *J. Assoc. Offic. Anal. Chem.*, 1986, **69**, 931.
[287] S. S.-C. Tai, N. Cargile, C. J. Barnes, and P. Kijak, *J. Assoc. Offic. Anal. Chem.*, 1990, **73**, 883.
[288] M. J. Lynch, D. M. Burnett, F. R. Mosher, M. E. Dimmock, and S. R. Bartolucci, *J. Assoc. Offic. Anal. Chem.*, 1986, **69**, 646.

be confirmed by transformation into the *cis*-isomers by high-intensity UV irradiation before HPLC. Recoveries of both drugs added at $2\,\mu g\,kg^{-1}$ were 60% and LoDs were $<0.5\,\mu g\,kg^{-1}$.

In a method[287] for pyrantel in pig liver, TAA was isolated after hydrolysis by ion exclusion chromatography and methylated for screening by GC–FID, where a co-extractive eluted very close to TAA methyl ester, and confirmation by GC–MS, where it was necessary to monitor relatively low mass ions. Detection limits, again inferred from blank tissue responses, were *ca.* 100 and 1000 $\mu g\,kg^{-1}$ for GC–MS and GC–FID respectively. Recovery at 1000 $\mu g\,kg^{-1}$ was *ca.* 60%.

2.17.9 Novobiocin

Moats and Leskinen[289] developed a chemical method for the drug in milk and cattle tissues. Sample treatment was similar to that used for virginiamycin:[290] blending with acidic phosphate and deproteinization with methanol. RP-HPLC analysis was by direct injection of large volumes of the deproteinized extract followed by gradient elution with 0.01 M phosphoric acid–acetonitrile–methanol and UV detection at 340 nm. C_{18} concentration columns varied in their breakthrough characteristics. Recoveries were uniformly 90–100% at 100–1000 $\mu g\,kg^{-1}$ and LoDs were in the region of $10\,\mu g\,kg^{-1}$.

2.17.10 Rafoxanide

Blanchflower and colleagues[291] monitored the drug in cattle and sheep tissue by homogenization with water, extraction into acetone–basic phosphate buffer, partition into petroleum ether, and RP-HPLC in acetonitrile–THF–0.1 M ammonium acetate (60:5:35), followed by thermospray MS. Selective ion monitoring using the molecular ion at m/z 626 permitted an LoD of 20 $\mu g\,kg^{-1}$. Confirmation could be achieved using the chlorine isotope ions but with loss of sensitivity. Recovery at 50 $\mu g\,kg^{-1}$ was 66% but averaged 80% at 100 and 200 $\mu g\,kg^{-1}$. The extraction procedure eliminated tissue-binding problems.

2.17.11 Virginiamycin

The drug is a complex of two factors and may contain variable amounts of the active components.[292] Some binding of virginiamycin to ovalbumin was observed.[292] The M_1 and S_1 factors have been determined[293] with LoDs of

[289] W. A. Moats and L. Leskinen, *J. Assoc. Offic. Anal. Chem.*, 1988, **71**, 776.

[290] W. A. Moats and L. Leskinen, *J. Agric. Food Chem.*, 1988, **36**, 1297.

[291] W. J. Blanchflower, D. G. Kennedy, and S. M. Taylor, *J. Liq. Chromatogr.*, 1990, **13**, 1595.

[292] D. E. Corpet, M. Baradat, and G. F. Bories, *J. Agric. Food Chem.*, 1988, **36**, 837.

[293] M. Nagase and K. Fukamachi, *Bunseki Kagaku*, 1987, **36**, 297 (*Chem. Abstr.*, 1987, **107**, 57 560).

100 and 10 $\mu g\,kg^{-1}$ respectively. The M_1 component was determined in pig tissue[290] by extraction and deproteinization as for novobiocin,[289] followed by defatting and solvent partition. RP-HPLC was carried out by on-line concentration and gradient elution. Detection was by UV at 254 nm and the LoD estimated as 10 $\mu g\,kg^{-1}$.

Acknowledgements

I wish to thank my colleagues G. Shearer, J. Carter, W. Farrington, G. Stubbings, and J. Tarbin from the Food Surveillance and Investigation (Veterinary Drugs) Department of the MAFF Food Science Laboratory, for their helpful comments during the preparation of this review and J. Tarbin for providing drug structures. I am also grateful to K. Woodward of the Veterinary Medicines Directorate for comments and for information on veterinary drugs licensed in the UK.

CHAPTER 9

Analysis of Pesticides at Low Levels in Drinking Water

KEITH M. MOORE

1 Introduction

The presence of pesticides in the aquatic environment has been known since the development of sensitive analytical methods for determining organochlorine insecticides back in the late 1950s.[1] A survey of organochlorine insecticides in selected raw and treated waters in England and Wales was reported in 1968.[2] Although organochlorine insecticides were detected during this survey it was concluded that the low levels found in drinking water would contribute only a very small portion of the total dietary intake of pesticides. Until recently there had been little monitoring of the concentrations of pesticides in drinking water, because the very low levels present were not considered to be adverse to health. However, in the 1980s with the impetus of the EC Drinking Water Directive, the monitoring of pesticides became more common. From 1 January 1990 the routine monitoring of pesticides in drinking water has been required of water undertakers (Water Supply plcs and companies) by English and Welsh law under the Water Supply (Water Quality) Regulations 1989,[3] made under the Water Act 1989.

The introduction of the Water Supply Regulations has provided a stimulus for the development of analytical methods which are sufficiently sensitive and accurate to meet the requirements of the regulations, and of methods that are cheaper and easier to use than previous analytical

[1] F.M. Middleton and J.J. Lichtenberg, *Ind. Eng. Chem.*, 1960, **52**, 99A.
[2] B.T. Croll, 'Pesticide Residues in Water-II. Levels of Organochlorine Insecticides in Selected Waters in England and Wales', Water Research Association Report TP 62, 1968.
[3] Water Supply (Water Quality) Regulations 1989, Statutory Instruments 1989 No. 1147, HMSO, London, 1989.

methods. Since the Water Supply Regulations and their supporting documents define the analytical requirements for drinking water in the UK, it is appropriate to review them in some detail before discussing their implications for analytical methodology.

2 The Regulations Applying to Pesticides in Drinking Water

2.1 The Development of Regulations

Prior to 1980 the view accepted in the UK was that of the World Health Organization, *i.e.* that drinking water generally makes only a minor contribution to the daily intake of pesticides, but that the contamination of water by pesticides should be prevented as far as possible.[4]

In 1980 the European Community adopted the Directive relating to the Quality of Water intended for Human Consumption (commonly known as the EC Drinking Water Directive).[5] There are sixty-two parameters (organoleptic, physical, chemical, and microbiological) specified in the Directive. These include parameter 55, 'Pesticides and Related Products', which are defined as insecticides, herbicides, fungicides, polychlorinated biphenyls (PCBs), and polychlorinated terphenyls (PCTs). However, this definition is not precise and it is unclear whether other classes of pesticide (*e.g.* rodenticides or even growth regulators) and metabolites of pesticides are included. Maximum admissible concentrations (MACs) and occasionally guide levels were set for the parameter 55 compounds. The MACs for individual and total pesticides are 0.1 and 0.5 $\mu g l^{-1}$ respectively. It has been claimed that the MACs for pesticides were based on the detection limit for an analytical method for organochlorine insecticides at the time the Directive was formulated.[6] In general, it seems the rationale behind parameter 55 was that pesticides should not be present in drinking water.[7] The MACs were set by the Directive but insufficient guidance was given on the analytical performance required.

The Directive was implemented in England and Wales in 1982 by the issue of a circular by the Department of the Environment (DoE) and the Welsh Office.[8] However, two reservations concerning the MACs for parameter 55 were expressed in a Water Guidance letter issued by the DoE and the Welsh Office in 1986.[9] First, the MACs were not set using a toxicological basis, and secondly it was not possible to monitor compliance for many of the pesticides since analytical methods with sufficiently low

---

[4] European Standards for Drinking Water, WHO, 1970.
[5] EC Directive 80/778/EEC. The Quality of Water Intended for Human Consumption. *Off. J. Eur. Commun.*, No. L229/11-29.
[6] B.G. Johnen, *Pestic. Outlook*, 1989, **1**, 9.
[7] G. Premazzi, G. Chiaudani, and G. Ziglio, 'Scientific Assessment of EC Standards for Drinking Water Quality', Commission of the European Communities, Luxembourg, 1989.
[8] Joint Circular, Department of the Environment 20/82, Welsh Office 33/82.
[9] Water Guidance Letter, Department of the Environment/Welsh Office WP10/1986.

detection limits were not available. With the first of these reservations in mind the Water Guidance letter contained a list of pesticides and their associated health advisory values (*i.e.* maximum acceptable drinking water concentrations based on toxicological considerations) set by the DoE and Department of Health and Social Security. The letter also gave advice on the frequency of monitoring, available analytical methods, and the pesticides to monitor. Much of this was incorporated in the Water Supply Regulations and a subsequent Water Guidance letter published in 1989.[10]

2.2 The Water Supply Regulations

The Water Supply (Water Quality) Regulations 1989 were enacted under the Water Act 1989 among other reasons to enforce the EC Drinking Water Directive in English and Welsh Law. There are also similar regulations for Scotland (enacted in 1990) and Northern Ireland (currently being produced) to enforce the EC Drinking Water Directive throughout the UK. The Water Supply Regulations define pesticides and related products and set the same MACs as the EC Drinking Water Directive. In addition, the Water Regulations and supporting documents prescribe sampling points and sampling frequencies and specify the requirements for analytical performance.

2.3 The Requirements of the Water Supply Regulations

The requirements of the Water Supply Regulations in relation to pesticides in drinking water are as follows.

2.3.1 Sampling

Samples taken to monitor compliance with the Water Supply Regulations must be taken and stored such that they are: representative of the quality of water throughout the water supply zone at the time of sampling; not contaminated during sampling and analysis; kept under conditions which ensure that there is no material change in the concentration of pesticides; and kept for the minimum time possible (*i.e.* analysed as soon as practicable).

The number of samples to be taken is specified in relation to water supply zones. A water supply zone is an area designated by a water undertaker in which not more than 50000 people reside. The standard sampling frequency to be applied at consumers' taps is four samples per annum per water supply zone. However, if in three successive years the water samples taken contained less than $0.05 \ \mu\text{g}\,\text{l}^{-1}$ of a single pesticide, and less than $0.25 \ \mu\text{g}\,\text{l}^{-1}$ of total pesticides, and the water undertaker is sure that the concentrations will not rise in the subsequent year, then a

[10] Water Guidance Letter, Department of the Environment/Welsh Office WP18/1989.

reduced sampling frequency is allowed. The reduced sampling frequency is one sample per annum per water supply zone. If the MACs for single or total pesticides have been exceeded then an increased sampling frequency is required. The increased sampling frequency is either 12 or 24 samples per annum per water supply zone, depending on whether the water supply zone supplies more than 35 000 people or 7000 m^3 d^{-1}.

A water undertaker may sample from a point other than a consumer's tap, if the Secretary of State can be satisfied that analysis of a sample from this point will produce data which are unlikely to differ from data produced by analysis of a sample from a consumer's tap. Different sampling frequencies apply to these sampling points.

2.3.2 *Analysis*

The Water Supply Regulations state that samples should be analysed as soon as possible after sampling, by or under the supervision of a person competent to perform that task, using suitable equipment, and using analytical systems and methods which are capable of establishing (within acceptable limits of deviation and detection) whether the MACs have been exceeded. Also, the laboratory at which the samples are analysed should have a system of quality control that is subjected from time to time to checking by a person who is not under the control of the laboratory or the water undertaker and is approved for that purpose by the Secretary of State.

The document 'Guidance on Safeguarding the Quality of Public Water Supplies'[11] (referred to here as the Guidance Notes) gives further advice on what is required of analytical methods. This states that the analytical method should determine the total parameter concerned (*i.e.* dissolved, colloidal, and particulate-associated pesticide, plus pesticide in any other chemical form, *e.g.* both ester and free acid forms of phenoxy-acid herbicides). It also states that for most parameters covered in the Water Supply Regulations the performance required of analytical systems is:

(i) A maximum tolerable total error of individual results not exceeding one tenth of a prescribed concentration or 20% of the result, whichever is the greater;

(ii) A maximum tolerable total standard deviation of individual results not exceeding one fortieth of the prescribed concentration or 5% of the result, whichever is the greater;

(iii) A maximum tolerable systematic error (or bias) of individual results not exceeding one twentieth of the prescribed concentration or 10% of the result, whichever is the greater;

(iv) A limit of detection (4.645 times the within-batch standard deviation

---

[11] Guidance on Safeguarding the Quality of Public Water Supplies, Department of the Environment/Welsh Office, HMSO, 1989.

of results for blanks) equal to (or presumably, if possible, less than) one tenth of the prescribed concentration;

(v) A recovery not significantly less than 95% or significantly greater than 105%.

The definitions for terms such as total error and systematic error are given by Hunt and Wilson.[12]

Table 1 shows what these performance requirements would mean in the case of individual pesticides. What the requirements would mean in terms of the total pesticide parameter (defined in the guidance notes as the sum of the detected concentrations of individual pesticides) is more difficult to define, since it depends on the number of pesticides monitored. However, in the guidance notes it is recognized that the level of performance in Table 1 is not possible for many pesticides using currently available analytical methods. Therefore the guidance is that the best currently available analytical methods should be used. The methods published by the Standing Committee of Analysts are given as examples of these.

Table 1 *Ideal performance characteristics required of analytical methods*

Performance characteristic	Values of performance characteristic allowed	
	Analytical result $\leqslant 0.05\ \mu g\,l^{-1}$	Analytical result $>0.05\ \mu g\,l^{-1}$
Maximum tolerable total error	$0.01\ \mu g\,l^{-1}$	20% of result
Maximum tolerable total standard deviation	$0.0025\ \mu g\,l^{-1}$	5% of result
Maximum tolerable systematic error	$0.005\ \mu g\,l^{-1}$	10% of result
Limit of detection	$0.01\ \mu g\,l^{-1}$	$0.01\ \mu g\,l^{-1}$
Recovery	>95% but <105%	>95% but <105%

There are a large number of pesticides used both agriculturally and non-agriculturally in the UK and consequently it is not feasible to monitor all of them. Guidance is therefore needed on which should be monitored. The Guidance Notes state that each water undertaker is required to develop a monitoring strategy for pesticides based on the likely risk of particular pesticides being present in the water source serving the zone. In developing a monitoring strategy water undertakers are advised to assess, as far as practicable, which pesticides are used in significant amounts within the catchment area, and whether any of these pesticides are likely to reach a water source in a catchment, based on their properties and method of use and on local catchment knowledge. The Water Guidance letter WP

[12] D. T. E. Hunt and A. L. Wilson, 'The Chemical Analysis of Water—General Principles and Techniques', 2nd Edn., The Royal Society of Chemistry, London, 1986.

18/1989[10] gives information on the properties and usage rates of some pesticides likely to enter water, to help the water undertaker develop a monitoring strategy.

2.3.3 Analytical Quality Control

The importance of analytical quality control for checking and helping to improve the performance of analytical systems is recognized in the Guidance Notes.[11] For internal quality control, laboratories are expected to carry out the following procedures:

(i) Use an analytical method capable of achieving the required accuracy, and which is written down unambiguously and in sufficient detail.

(ii) Estimate the within-laboratory total standard deviation of individual analytical results for blanks, standard solutions, samples, and spiked samples having concentrations over the range of interest, over at least five batches on five separate days. Each estimate of total standard deviation should have at least ten degrees of freedom. The recovery of added spikes of the pesticide from typical samples should also be assessed to check certain sources of bias.

(iii) Set up and maintain a quality control chart to check the routine performance of the analytical system. As a minimum the control analysis should be of a standard solution with a concentration close to the prescribed concentration, *i.e.* $0.1 \mu g l^{-1}$ in the case of pesticides.

Laboratories are also expected to participate in external quality control schemes where available, which involve the distribution of check samples. Such a check sample quality control scheme is available for a variety of pesticides from the WRc Service Aquacheck. Some results obtained by Aquacheck are presented in Table 2. They show that there can be wide variations in the accuracy of analytical results when analysing pesticides at low levels in water. Some analytical results were in error by more than 100%. Other inter-laboratory studies have also shown that some pesticide analytical results can be very inaccurate,[13] and therefore it is important to have good internal quality control procedures.

3   Analytical Methods for Pesticides in Water

The discovery of organochlorine insecticides in raw waters was made possible by the development of gas chromatography (GC) in the 1950s. Methods similar to those used in the 1950s employing solvent extraction and gas chromatography are still used, but there have been significant developments in analytical techniques in recent years.

---

[13] L.A. Norris, *Weed Sci.*, 1986, **34**, 485.

Table 2 *Summary of results from AQUACHECK pesticide samples*

| Pesticide | Spiked concentration ($\mu g\,l^{-1}$) | No. of labs reporting | % of laboratories reporting | |
			Values in error by $\geqslant$50% of spiked value	Values in error by $\geqslant$100% of spiked value
Endrin	0.034	27	15	0
Dieldrin	0.056	31	16	3
Aldrin	0.049	33	15	3
pp'-DDT	0.058	29	38	3
Lindane	0.044	33	12	0
Simazine	0.084	15	0	0
Atrazine	0.078	15	13	7
Propazine	0.031	16	25	13
Azinphos methyl	0.063	3	33	33
Dichlorvos	0.072	4	0	0
Fenitrothion	0.065	5	20	20
Malathion	0.041	8	63	25
Mevinphos	0.054	3	33	17
Chlorfenvinphos	0.026	6	33	17

3.1 Official Analytical Methods

The official source of analytical methods for pesticides in water in the UK is the Standing Committee of Analysts (SCA). The methods currently available were all published before the introduction of the Water Supply Regulations. However, some of the methods are capable of achieving detection limits below 0.1 $\mu g\,l^{-1}$. Table 3 lists current SCA methods for pesticides and their detection limits. There are also a large number of SCA analytical methods for pesticides in draft form at the time of writing this chapter. Although limits of detection below 0.1 $\mu g\,l^{-1}$ are possible with some SCA analytical methods, problems with interferences have been encountered, for example with the phenoxy acid-herbicides.[14] Interferences can be encountered as many existing SCA methods use non-specific detection techniques.

The current SCA analytical methods do not meet all the requirements for analytical performance described in Table 1. However, the Guidance Notes state that as current methods for pesticides cannot meet these requirements, the best currently available analytical methods (*e.g.* the SCA methods) should be used.

---

[14] A. Waggott, Proceedings of COST 641, Workshop held at Water Research Centre, Medmenham, UK, 23–24 November 1988, ed. B. Crathorne and G. Angeletti, Commission of the European Communities, Water Pollution Research Report II, 1989.

Table 3 *Currently available SCA methods for pesticides*

Pesticides covered	Reference	Detection limits quoted (μg l^{-1})
Organochlorine insecticides	a,b,c	0.003–0.015
Total PCBs	a,b,c	0.1
Organophosphorus insecticides	d,e	0.03–0.4
Chlorophenoxy acidic herbicides	f	0.004–0.1
Trichlorobenzoic acid	f	0.0005
Chlorophenols	f	0.02–0.2
Triazines	f	0.025
Glyphosate	f	0.08
Carbamates	g	0.02–0.29
Thiocarbamates	g	0.48 as CS$_2$
Urons	g	<5
Diquat & Paraquat	h	0.02
Pyrethrins	i	0.24
Permethrin	i	0.06

[a] Organochlorine Insecticides and PCBs in Water, Standing Committee of Analysts, HMSO, London, 1978.

[b] The Determination of Organochlorine Insecticides and PCBs in Sewage, Sludges, Muds and Fish 1978, in Water (an addition), Standing Committee of Analysts, HMSO, London, 1984.

[c] Chlorobenzenes in Water, Organochlorine Insecticides and PCBs in Turbid Waters, Halogenated Solvents and Related Compounds in Sewage Sludge and Waters, Standing Committee of Analysts, HMSO, London, 1985.

[d] Organophosphorus Pesticides in River and Drinking Water (tentative method), Standing Committee of Analysts, HMSO, London, 1980.

[e] Organophosphorus Pesticides in Sewage Sludge, in River and Drinking Water, Standing Committee of Analysts, HMSO, London, 1985.

[f] Chlorophenoxyacidic Herbicides, Trichlorobenzoic Acid, Chlorophenols, Triazines and Glyphosate in Waters, Standing Committee of Analysts, HMSO, London, 1985.

[g] The Determinaton of Carbamates, Thiocarbamates, Related Substances and Ureas in Water, Standing Committee of Analysts, HMSO, London, 1987.

[h] Determination of Diquat and Paraquat in River and Drinking Water, Spectrophotometric Methods (tentative), Standing Committee of Analysts, HMSO, London, 1987.

[i] Pyrethrins and Permethrin in Potable Waters by EC-GC, Standing Committee of Analysts, HMSO, London, 1981.

3.2   Developments in Analytical Methodology to Meet the Requirements of the Water Supply Regulations

The majority of analytical methods currently used are based on solvent extraction followed by chromatography and detection with a non-specific detector. There have been developments in both the extraction and detection steps of the conventional pesticide analytical methods, and some new approaches to pesticide analysis have recently emerged, involving automation and/or analytical methods using different principles (*e.g.* immunoassays).

3.2.1   Developments in the Extraction of Pesticides

A development which can improve the performance of the extraction step in pesticide analysis is the use of solid-phase extraction (SPE). This technique involves extraction of water samples using a solid sorbent material, such as amberlite XAD resin and, more recently, bonded silica sorbents in cartridges. For many pesticides SPE achieves better extraction efficiencies than solvent extraction.[15] SPE also offers other advantages over solvent extraction, including: greater sample throughput rate, ease of use, less use of potentially harmful solvents, potential for automation, and greater selectivity.

Bellar and Budde[15] reported that the precision was worse for SPE than for solvent extraction, but the pesticides for which precision was poor were those for which the recoveries were low. Probably if more suitable stationary phases were used for these pesticides, then better precision would be obtained.

3.2.2 Developments in the Detection of Pesticides

Analytical methods using non-specific detectors, such as GC–ECD, are satisfactory for pesticides in drinking water when monitoring compliance with toxicologically based water quality standards since such standards are invariably considerably higher than $0.1 \mu g l^{-1}$. However, interferences may be encountered when using these methods to monitor compliance with the $0.1 \mu g l^{-1}$ MAC. When an analytical method is being used to monitor one or two pesticides it is normally easy to choose a chromatographic column which separates the pesticides from the interferences. However, when larger numbers of pesticides are being analysed in one chromatographic run (*i.e.* in a multi-residue method) this becomes more difficult. The problem encountered with interferences when using a nitrogen phosphorus detector (NPD) for analysing 'dirtier' water is illustrated in Figure 1.

The problem of interferences can be solved in two ways. One approach is to eliminate the interferences using a clean-up technique. This approach has been used for the analysis of triazine herbicides[16] and phenoxy-acid herbicides.[17] An ion-exchange clean-up technique was used in both of these methods and it appears to have been successful, as can be seen from the HPLC chromatograms shown in the respective papers. Other types of clean-up can be used, such as normal phase column chromatography on Florisil, silica gel, or alumina. The disadvantage of this approach is that the pesticides to be determined must be chemically similar to each other but different from the interferences. This means that the number of pesticides that can be determined by any one method is limited.

Another approach for avoiding interference is to use a highly selective detector such as a mass spectrometer. This approach is very successful as

[15] T.A. Bellar and W.L. Budde, *Anal. Chem.*, 1988, **60**, 2076.
[16] M. Battista, A. DiCorcia, and M. Marchetti, *Anal. Chem.*, 1989, **61**, 935.
[17] A. DiCorcia, M. Marchetti, and R. Samperi, *Anal. Chem.*, 1989, **61**, 1363.

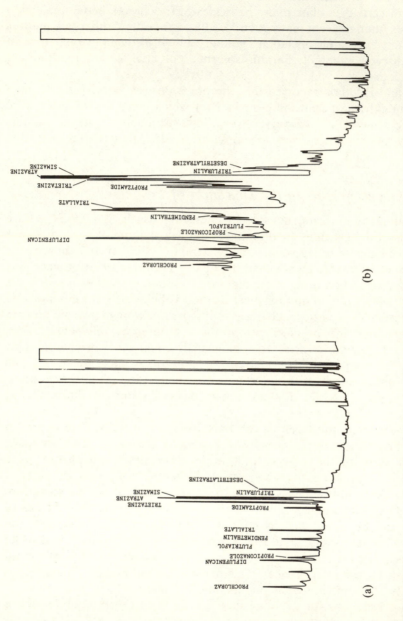

Figure 1 *GC–NPD chromatograms of (a) borehole derived drinking water spiked at 0.1 µg l⁻¹ with various nitrogen-containing pesticides; (b) river water spiked at 0.1 µg l⁻¹ with various nitrogen containing pesticides: pH 8, dichloromethane extracts reconstituted in 100 µl of ethyl acetate; 1 µl on-column injection onto 30 m DB5 (J&W) column held at 60 °C for 3 min, then programmed to 300 °C at 8 °C min⁻¹*

can be seen in the gas chromatograph–mass spectrometer (GC–MS) trace in Figure 2. Mass spectrometry is a selective and sensitive detection technique which, with the different ionization modes now available (thermospray, chemical ionization, *etc.*), can be used to determine the majority of pesticides. The disadvantage of mass spectrometry is that it is more costly than other detectors and consequently it is not so readily available (although most water undertakers have access to one). However, when less costly non-specific detectors are used it is advisable to confirm significant results using mass spectrometry.[18]

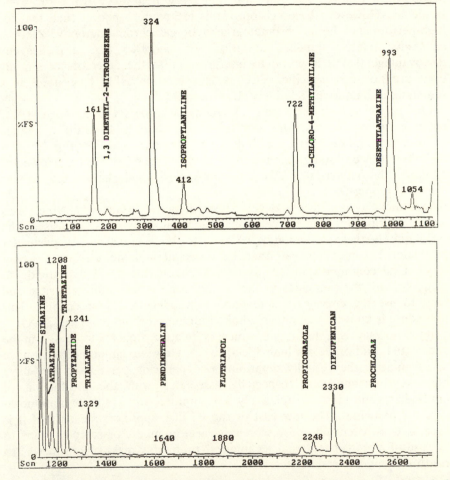

Figure 2 *GC-MS selected ion monitoring total ion current chromatogram of borehole water spiked at 0.1 µg l⁻¹ with various nitrogen containing pesticides: pH 8, C₁₈ solid-phase extract reconstituted in 100 µl of ethyl acetate; 1 µl on-column injection onto 30 m DB1 (J&W) column held at 60 °C for 3 min, then programmed to 300 °C at 8 °C min⁻¹*

[18] U. Oehmichen, F. Karrenbrock, and K. Haberer, *Gewaesser. Wasser Abwasser*, 1989, **106**, 110.

3.2.3 Developments in Producing Lower Cost and Automated Methods

The Water Supply Regulations require a considerable number of pesticide analyses to be carried out, and so there is a need for low cost and automated methods. There have been various approaches to this including thin-layer chromatography (TLC), on-line solid-phase extraction, and immunoassay.

Thin-layer chromatography (TLC) is a cheap semi-automatable technique which can achieve detection limits just below $0.1 \mu g l^{-1}$ for some pesticides. However, the precision of this technique is poorer than that for high-performance liquid chromatography or gas chromatography; Zietz *et al.*[19] give relative standard deviations of 15–20%. TLC also has the disadvantage that it is prone to interferences as the detectors commonly used are not very specific. TLC is therefore probably best used as a screening technique for relatively clean samples.

On-line solid-phase extraction is a technique using small cartridge HPLC columns to concentrate the pesticides and then a separate analytical column for on-line HPLC analysis (*e.g.* Hamann and Kettrup[20]). This technique can be totally automated and achieve good precision. However, it limits the analytical method to pesticides which have very similar chemical properties.

Immunoassay is a recent development in pesticide analysis, which uses different principles from conventional chemical analysis. The basis of the technique is that an antibody is used to bind selectively a particular pesticide. The pesticide can then be detected in a number of different ways. One common approach is to use an enzyme-labelled pesticide to compete with the pesticides in the sample for the free antibody sites and then to use the enzyme in a reaction to produce a colour change. This procedure is known as an enzyme-linked immunosorbent assay (ELISA).

Immunoassay is a technique which is relatively cheap to use, can be automated, and often has high specificity.[21] However, immunoassays have disadvantages: they usually determine one pesticide per assay, they have high development costs (especially associated with the production of antibodies), and there are relatively few commercially available immunoassays for pesticides. The low cost of the ELISA approach means that it is best used as a technique for screening large numbers of samples prior to the use of more conventional techniques (preferably mass spectrometry) to confirm significant results.

4 Conclusions

The Water Supply Regulations are stringent with respect to the perform-

[19] E. Zietz, I. Ricker, and G. Arendt, *Gewaesser. Wasser Abwasser*, 1989, **106**, 136.
[20] R. Hamann and A. Kettrup, *Chemosphere*, 1987, **16**, 527.
[21] F. Jung *et al.*, *Pestic. Sci.*, 1989, **26**, 303.

ance required of analytical methods for monitoring compliance with the MACs for pesticides. For many pesticides this performance is difficult to achieve using currently available analytical methods. Inter-laboratory quality control studies have shown that there can be large inaccuracies in pesticide analytical data. It is therefore important that analytical methods are thoroughly tested to assess their performance, and quality control procedures are used to check that the required quality of analytical data is maintained.

Some currently available analytical methods which employ non-specific detectors can have problems with interferences. When using these methods it is advisable to confirm positive results using mass spectrometry.

In principle, the Water Supply Regulations require a considerable number of pesticide analyses to be carried out. Some recent developments in pesticide analysis which help make pesticide analyses cheaper or faster are solid-phase extraction and immunoassay. Solid-phase extraction is becoming more widely used because among its advantages it is quicker, easier to automate, and uses significantly smaller volumes of potentially harmful solvents than solvent extraction. Immunoassay appears to be a promising technique for the screening of large numbers of samples because once developed it is cheap to run, and quick and easy to automate, but it is advisable to confirm significant results using a conventional analytical technique, preferably based on mass spectrometry.

Acknowledgements

I would like to acknowledge Mr. M. Gardner for permission to use the Aquacheck results and Mr. M. Fielding, Mrs. I. Wilson, Mrs. K. Wilson, and Miss S. Gibby for technical assistance.

CHAPTER 10

Unwanted Flavours in Foods

J. DAVID HENSHALL

1 Introduction

In today's complex food chain starting on the farm, the fishing boat, or even deep in the tropical jungle and ending on the consumer's dining table, there is a prime requirement for the processor to provide a product of safe and consistent quality which has strong consumer appeal. Three main attributes are carefully controlled, these being colour, texture, and flavour.

The raw materials of the food are largely of natural origin and are subject to variation in their characteristics. The art or skill of the food scientist and technologist is to control the formulation and processing of raw material to produce the required products. However, things do go wrong!

Flavour problems in a wide range of processed foods, including 'fresh produce', are increasing. The incidence of these outbreaks is still very low when the total volume of food passing through the distribution system is considered and the question which concerns those who supply high quality products is 'Why is this so?'. Either there is (i) a real problem, (ii) the consumer's expectations are higher than previously when variations in quality attributes were more tolerated, or (iii) consumers are more prepared to complain. Whatever the cause, there has been increased interest in the analysis of off-flavours and taints.

Off-flavours in food have always been with us and historically these have been caused by microbial or fungal contamination, or oxidative deterioration. Sometimes these strong flavours have been welcomed and enhanced and over many years have resulted in products which are now considered to be delicacies – the same flavour in another food is considered to be objectionable. In earlier times when food storage was a primitive 'process', these off-flavours were often disguised by the use of spices. The scene today is different – the consumer expects, and rightly so, food which is safe to eat and exhibiting good organoleptic properties. Therefore, when one of the major attributes such as flavour is not acceptable or just plain objectionable, the food supplier wishes to know what happened and why.

191

Food processors are often warned about a taint problem by consumers returning products through a retailer or by their making direct approaches to the company. Often it is possible to gain an idea of the severity of the problem by the numbers of complaints and the written comments which are received. Sometimes the geographical area from which the complaint arises can provide useful information if there is more than one notification – perhaps contamination occurred *en route* to the area or while the product was stored in a particular distribution centre. Even so, it is unwise to rely on consumers alone for a precise characterization or description of the taint – a trained tasting panel should be used. It is of great importance to use an expert panel to detect and describe the taint because for any chemical, in a human population, there will exist a wide range of sensitivities. Therefore in a panel, there should be people with sensitivities greater than those of the population mean.

In the past thirty years, many new synthetic chemicals have been introduced into the environment and some of these have entered the food chain and can be shown to occur in raw materials either as non-degraded compounds or metabolites. Today, both processed and fresh foods are affected by flavours variously described as earthy, catty, solvent-like, painty, mouldy, faecal, metallic, or cardboardy. These flavours may be introduced into the food or the packaging material by absorption from many sources including the environment in which food has been stored, grown, or processed. The environment in this context means soil, air, and water. Many of these compounds are man-made and it is these which cause some of the more severe flavour problems and about which precautions can be taken to reduce the risk of food contamination.

The compounds of interest in identifying a food as well as providing subtle differences to products of similar origin are the volatile components. Foods contain many hundreds of such compounds which either singly, or in combination, define precisely the identity of an acceptable food product and contribute to human or animal consumer preferences between different brands, products, or varieties. It is also the volatile compounds which are almost always responsible for the off-flavours which arise from contamination or chemical, biochemical, or microbiological transformation.

2   Analytical Methods

Mixtures of food flavour volatiles are invariably separated by gas chromatography (GC), almost always followed by mass spectrometry for the identification of particular compounds. Before analysis, however, there is the problem of extraction of the flavour components from the food matrix in a suitable form and concentration for GC analysis. The extraction stages usually employ solvent extraction, distillation, trapping of headspace volatiles on to an absorbent, or a combination of these techniques. The range of these combinations in the published literature is large, and the problem

is that each process is likely to produce different flavour profiles on analysis.

A typical GC trace of the volatile components from yeast is shown in Figure 1 and the complexity of the mixture can be seen.[1] The tainting compound in the sample was indole, eluting at 32.5 minutes. This was confirmed by mass spectrometry; the mass spectrum is shown in Figure 2. This illustrative sample is unusual in that the quantity of contaminant is large. A useful technique to narrow the selection of a peak and to relate it to the occurrence of a particular flavour is to sniff the effluent gas from the chromatograph.[2] The processes of extraction and concentration of flavour components have been the subject of many studies and it is known that each technique has advantages and disadvantages. A good example is that of sample/solvent co-distillation; the heating process may cause artefact formation and so the components of the volatile mixture should be compared with the results obtained by another technique, for example closed loop stripping.[3]

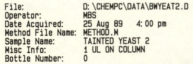

```
File:              D:\CHEMPC\DATA\BWYEAT2.D
Operator:          MBS
Date Acquired:     25 Aug 89   4:00 pm
Method File Name:  METHOD.M
Sample Name:       TAINTED YEAST 2
Misc Info:         1 UL ON COLUMN
Bottle Number:     0
```

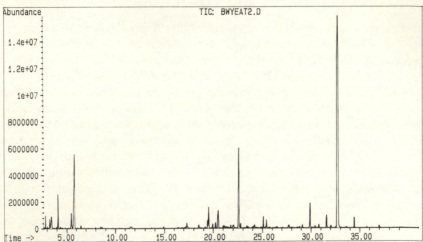

Figure 1 *GC trace of volatiles from a tainted sample of yeast*

[1] M. B. Springett, personal communication, 1990.
[2] N. M. Griffiths and D. G. Land, *Chem. Ind. (London)*, 1973, 904.
[3] K. Grob, Jr. and F. Zurcher, *J. Chromatogr.* 1976, **117**, 285.

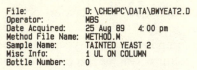

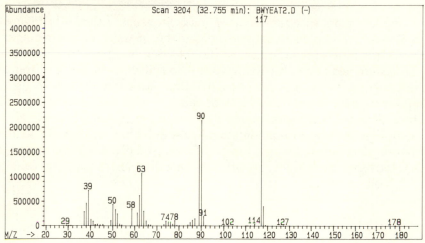

Figure 2 *Mass spectrum of indole, originating from a tainted sample of yeast*

3   Sources of Taints and Off-flavours

Taint is the term which is used to describe alien flavours in foods which come from outside sources and are perceived by a taster as being unpleasant. Generally speaking, the chemicals which cause taint are present in very low concentrations and are detected because of the low threshold values. These are nearly always at the parts per billion (μg kg^{-1}) level. Two consequences arise therefore. The first is that quantitation is difficult and requires sophisticated techniques and instrumentation to extract such small quantities from a complex and variable matrix. Secondly, in practice in instances where tainting occurs toxicity is not usually a problem for the consumer. (The latter observation obviously excludes malicious contamination which is quite another matter.) 'Off-flavour' is another term used to describe food that does not have the expected or required flavour. This term should be used to describe those flavours which arise from microbiological activity or deteriorative processes such as oxidation and rancidity. A combination of sensory and chemical analysis should be used to trace the source of a problem – the human used as a detector is often more sensitive than chemical analysis, but because tainting chemicals behave differently in different foods, identification by chemical methods should always be undertaken.

Since the publication of the seminal article by Goldenberg and Matheson,[4] much experience has been gained by food analysts in this most

interesting area of food chemistry. A further survey was that written by Whitfield,[5] and an important work edited by Charalambous[6] has recently been published. The number of taint problems and descriptions of techniques referred to in these references indicates the magnitude and complexity of describing definitively the cause of an outbreak of off-flavours in a foodstuff. Coupled with this is the knowledge that, following the identification of a problem, the solution is always required urgently because the distributor and manufacturer have to put recall procedures very quickly into effect. It may be that a product has been tampered with and the offending chemical or chemicals may be toxic – in any case, responsible companies want an urgent answer in order that the problem is not repeated.

Various descriptions of off-flavours or taints have been defined. Experience has shown that the chlorophenols, such as 6-chloro-2-methylphenol, are responsible for disinfectant type flavours while the chloroanisoles cause mouldy and musty effects.[7,8] The isomer of the phenol is important; for example, 4-chloro-2-methylphenol has a taste threshold of $3 \, mg \, kg^{-1}$ in biscuits and $0.12 \, mg \, l^{-1}$ in water. For 6-chloro-2-methylphenol, the values are $5 \times 10^{-5} \, mg \, kg^{-1}$ in biscuits and $8 \times 10^{-5} \, mg \, kg^{-1}$ in water. To produce a disinfectant taint in biscuits requires $1 \times 10^{-4} \, mg \, kg^{-1}$ of the latter compound. Goldenberg[4] stated that when biscuits were exposed for 10 min to air at 300 yards from a herbicide factory, $0.08 \, mg \, kg^{-1}$ of mixed chloro-2-methylphenols could be found in the biscuits, resulting in what was described as a 'gross contamination producing an offensive smell and taste'.

Very often a cause of food contamination by chlorophenols is from the water which is used in processing, either as an ingredient or as a processing aid. Formation of chlorophenols can occur by the use of heavily chlorinated water reacting with the food or by chlorinating certain types of mains water, for example those derived from peaty sources. In the former case, chlorine can react spontaneously with simple phenols, while in the latter the problem has occurred industrially when the chlorinated water has been used for steam generation in high-pressure boilers. A description often used in food off-flavour terminology is 'musty' and this sometimes arises from water. Maga[9] identified several molecules; geosmin (*trans*-1,10-dimethyl-*trans*-9-decalol), 2-methylborneol (2-methylisoborneol), and an alkoxypyrazine (2-methoxy-3-isopropylpyrazine) as causing this effect. The odour-producing properties of these compounds vary according to the concentration. They are probably due to algal blooms in the water and indicate the importance of monitoring raw water quality in food processing operations,

[4] N. Goldenberg and H. R. Matheson, *Chem. Ind. (London)*, 1975, 551.
[5] F. B. Whitfield, *CSIRO Food Res. Q.*, 1983, **43**, 96.
[6] 'Flavours and Off-flavours', ed. G. Charalambous, Elsevier, Amsterdam, 1990.
[7] R. L. S. Patterson, *Chem. Ind. (London)*, 1972, 609.
[8] C. Engel, A. P. de Groot, and C. Weurman, *Science*, 1966, 270.
[9] J. A. Maga, *Food Rev. Int.*, 1987, **3** (3), 269.

expecially during warm weather. The use of algaecides to prevent algal growth is not to be recommended. Some have been shown to contain 2,4,6-trichlorophenol and 2,3,4,6-tetrachlorophenol, and Whitfield[5] cited a case where water treated in such a manner leaked from a refrigeration plant into process water and caused difficulties.

Until some years ago, the use of phenolic resins or paints in food processing operations gave rise to problems – not only of phenolic taints but also hydrocarbon-type flavours. It is now common practice – if not universal – to test all materials which come into contact with food or which may be used in the construction of food factories and handling facilities for their liability to taint food, and therefore contamination from this source is now, happily, rare.

Pesticides have been implicated, and proved, to be the source of phenolic taints.[5] Episodes from this source still occur and problems can be found when treated timber comes into contact with or in proximity to food. The wood pulp from which packing cases are made, wooden pallets, paper sacks, and adhesives can all contain residual fungicides, and contamination from these sources may occur; the offending flavour is usually phenolic. An interesting account of a sporadic taint problem in soft drinks by contamination from this source is given by Whitfield.[5] During two consecutive summers, an Australian soft drink manufacturer suffered from random outbreaks of a medicinal taint in his products. It was shown that the off-flavour was not in the product prior to canning and that the problem occurred only in a few cans at any one time. The compound causing the off-flavour was not identified in the early stages of the investigation. The source of the taint was eventually identified as arising from an agrochemical factory nearby. It was believed that the cause was 6-chloro-2-methylphenol which was carried by the wind to the soft drinks plant and that the material was absorbed by the can lacquer where cans were partially exposed at the edge of the pallets. The cans could be reclaimed by restoving but the cardboard spacers on the pallets were destroyed because they could have provided a continuing source of taint. At the time, 6-chloro-2-methylphenol and 4-chloro-2-methylphenol were significant impurities in the herbicide 4-chloro-2-phenoxyacetic acid which the agrochemical plant happened to be producing.

Baigrie[10] describes an instance where an antiseptic type flavour was thought to be caused by a chlorophenol; however, some members of the tasting panel believed that the off-flavour did not have all the attributes of a true chlorophenol off-flavour – subsequent analysis showed that the compound responsible was N,N-dimethylaniline.

Elegant studies by Curtis *et al.*[11] showed that *Aspergillus* and *Penicillium* species can methylate chlorophenols to produce chloroanisoles and these were ingested by poultry which then exhibit intense off-flavours when

[10] B. D. Baigrie *Food Manuf.*, 1988, August, 21.
[11] R. F. Curtis, C. Dennis, J.M. Gee, N. M. Griffiths, D. G. Land, J. L. Peel, and D.G. Robinson, *J. Sci. Food Agric.*, 1974, **25**, 811.

cooked. A puzzling problem of a mouldy off-flavour in meat exported from Australia revealed that the problem was caused by condensate in chiller rooms flooding insulation material which contained a small amount of phenol. The chiller rooms were sterilized by means of chlorine, which reacted with the phenol, which was in turn methylated by fungus present in the building material. Forced air circulation then transported the compound throughout the building. The problem was solved by closing down the chiller complex and rebuilding.[12]

To indicate that the problems are still in existence today, there is much work on 'cork taints' in alcoholic drinks, *e.g.* Cognac[13] and wine.[14–16] Some taints are due to treatment of cork trees with fungicides and others are due to contamination of damp warehouses by fungi. Careful analysis is required to identify the true cause of the problem.

Other types of off-flavour can be caused by printing inks and paint solvents containing styrene and naphthalene. A particularly nasty taint can be caused by compounds which contain ester or ketonic functional groups such as methyl vinyl ketone.[17,18] The ketones react with sulphydryl groups in the food causing catty odours which are particularly noticeable and offensive. Outbreaks have occurred in a wide range of foods such as carrots (solvent residues from pesticides), pork luncheon meat (residual solvent in can lacquers), rice (printing ink on paper sacks), and cakes (airborne contamination produced by the interaction of ketones and sulphur compounds in a chemical factory 20 miles away from the bakery).[4]

Swoboda and Peers[19] have attributed the occurrence of cardboard flavours in foods to the presence of saturated or unsaturated aldehydes and metallic flavours to unsaturated ketones. These compounds can occur in foods by oxidative processes. It is interesting to note that some cardboards can contain up to 2% by weight of non-polar lipid – in other words, the 'cardboardy' description is derived from the fact that paper has undergone oxidation.

4 Case Studies

In recent years a wide range of off-flavours and taints have been identified in a variety of foods (Table 1). Some of these have involved extensive detective work to elucidate the various causes. Successes in identifying these causes have been the result of close collaboration between food

[12] CSIRO Division of Food Research, Report of Research 1981–82, p. 25.
[13] R. Cantagrel and J. P. Vidal, in 'Flavours and Off-flavours', ed. G. Charalambous, Elsevier, Amsterdam, 1990, p. 139.
[14] H. Tanner, C. Zanier, and G. Wurdig, *Schweiz. Z. Obst. Weinbau*, 1981, **117**, 752.
[15] H. Tanner, C. Zanier, and H. R. Buser, *Schweiz. Z. Obst. Weinbau*, 1981, **117**, 97.
[16] P. Dubois and J. Rigaud, *Vignes et Vins*, 1981, **301**, 48.
[17] T. J. P. Pearce, J. M. Peacock, F. Aylward, and D. R. Haisman, *Chem. Ind. (London)*, 1967, 1562.
[18] F. Aylward, G. Coleman, and D. R. Haisman, *Chem. Ind. (London)*, 1967, 1563.
[19] P. A. T. Swoboda and K. E. Peers, *J. Sci. Food Agric.*, 1977, **28**, 1019.

technologists, analytical chemists, and sensory panels. Others defied solution; only theories could be put forward.

Table 1 *Food tainting problems – some products and the causes*

Disinfectants	*Sulphur taints*
Tomatoes – fresh	Cider
Coated fish products	Wine
Tomato juice	
	Bacterial/fungal
Chlorophenols	Orange juice – acetoin
Milk	Fried potatoes – ethyl butyrate
Water	Frozen pastry – ethyl butyrate
Cardboard packaging	
Sugar solutions	*'Catty' taints*
Grain	Meat products
Fish paste	
Canned beans	*Enzyme reactions*
Red kidney beans	Pizzas
Chloroanisoles	*Miscellaneous/unidentified sources*
Strawberry jam	Milk – faeces odour – methylindole
Cocoa butter	Apples – naphthalene
	Apples – blue–green discoloration
Solvents/lubricants	plus a metallic flavour
Coated fish products	Custard – cabbagy off-flavour
Biscuits	Corned beef – musty flavour
Pineapple	Ice-cream – rancid palm oil
Lamb (also contained naphthalene)	Poultry – contamination from plastic trays
Rice	Wine – contamination by non-food grade
Corned beef	plastic tube

The most common problems, in terms of numbers of outbreaks, have been due to chlorophenols and chloroanisoles. The means of contamination have sometimes been quite unusual – for example, greenhouse-grown fresh tomatoes absorbed chlorophenols from traces of these compounds brought into the house from a sanitizing footbath outside the entrance. A coated fish product was tainted by a disinfectant used on board a trawler – the catch by this boat was purchased by a processor and mixed with other lots. Eventually the whole day's production in a fish processing factory was contaminated.

Process water caused chlorophenol problems – the mains supply to a factory was changed, the new supply being abstracted from a different source and containing phenols derived from humus. A similar problem occurred in a pea cannery. These problems came to light when contaminated product was detected during quality control operations. Others were caused by residual cleaning materials following clean-in-place operations or by too high a level of free chlorine in water used for fluming.

Chloroanisoles were detected in cocoa butter. The origin of these was suspected to be fungal methylation of chlorophenols during fermentation of the cocoa beans in the source country. The reason why chlorophenols were present in the first place is unknown as the field history of the crop could not be traced. In the case of the strawberry jam, chloroanisoles were detected in cardboard boxes used for transporting the fresh fruit.

Solvent or lubricant taints can usually be explained by accidental contamination during processing or storage, or by storage of food near to products releasing tainting compounds. Departures from good manufacturing practice are often the cause of this type of problem, for example where containers in a factory are used for non-food purposes. The tainting of frozen desserts was caused by solvent release from mastic used for sealing around holes for pipes running through a wall. This tainting process occurred at −40 °C in a freezer store.

Similarly, bacterial or fungal taints are usually attributable to abuse of the food. The fried potatoes in Table 1 had a cheesy–sour flavour caused by *Pseudomonas* spp. producing ethyl butyrate, as did the frozen pastry. The orange juice contained acetoin, either as a result of over-processing or bacterial contamination. The colour of the product favoured the over-processing theory and no significant number of dead bacterial cells were found. A problem which can occur in a warm pea season is that of 'vining flavour', where mechanically harvested peas are held for too long before entering the factory and being processed. This is a combination of chemical (mainly) and microbiological deterioration.

A most interesting problem presented itself with frozen pizzas, occuring spasmodically during several days production. The pizzas were garnished with fresh sliced capsicums and it was eventually established that, in some consignments of mozzarella cheese, a compound or compounds reacted with some component of the peppers to cause the effect. No effect could be produced with blanched peppers, and efforts to duplicate the reaction using other batches of cheese and peppers failed. The effect was assumed to be enzyme moderated because the intensity of the effect could be changed at different storage temperatures. In view of the limited time and quantity of material available, no analytical work was carried out but the basic cause of the off-flavour was established beyond reasonable doubt.

Several examples of off-flavour problems have proved to be insoluble. Why, for example, did frozen bramley apple turn blue–green on thawing and simultaneously develop a metallic off-flavour? What caused a custard product to taste of cabbage? Where did fresh apples become contaminated with high levels of naphthalene? In these cases the level of occurrence of the problem was very small, but by those who suffered each episode was considered to be very important because somewhere, something had gone wrong and for an ethical industry much effort had to be expended to characterize the problem in order to prevent its recurrence. In these problems it was not possible to find the root cause of the contamination.

5 Prevention

What can the food manufacturer do to reduce the possibility of tainting chemicals entering the product? The number of problems encountered by the industry is growing despite the increased knowledge base that is being accumulated. There are more sources of chemicals being recognized as causing taint in food. Some possible reasons may be the use of new packaging materials and the current concerns about migration of components into the food, the increased expectations of the consumer, the increased use of water which may carry traces of industrial chemicals, problems of air and soil pollution, and finally, perhaps, the use of bulk transport containers which have been used to carry odoriferous materials.

To combat these, the food processor must check every aspect of the process and ensure that a good quality control system based on BS5750/ISO9000/EN29000 is implemented. Many companies in the food and allied industries are now doing this. In the case of possible air-borne contamination, food technologists should be aware of likely problems from nearby chemical factories; new food factories and warehouses should be sited well away from sources likely to produce contamination. In the case of packaging, careful attention should be paid to specifications and these should be checked regularly after the initial drawing-up. Water used for processing or as an ingredient should be monitored to ensure that pollution has not occurred. Paints and flooring materials should be allowed to cure before food processing takes place in a building which has been erected or refurbished. Process plant and buildings should be designed and maintained so that pockets of soil where bacteria and fungi may multiply are not present. Agricultural chemicals used on crops should be screened for their propensity to produce taints or off-flavours. Sub-contractors and suppliers should be advised of the problem of taint and warned to take precautions.

If all these points are taken into consideration, then the likelihood of a taint problem occurring will be lessened and the question 'Who is to blame?' will not be heard as often.

Subject Index

Acceptable daily intake, 66
  monomers and additives, 76
  pesticides, 67
Adverse Reaction Scheme, 102
Advisory Committee on Pesticides, 63
Aflatoxins, 18
Animal Test Certificate, 102

BIBRA/British Plastics Federation
  Code of Practice, 74
Bracken, 14
  carcinogenic properties, 5, 14
  illudane sesquiterpenes, 14
Brassicas, 7
  and glucosinolates, 10

Cassava, 6, 17
Celery, 6
  and psoralens, 6
α-Chaconine, 8, 13
Chlorophenols, 23, 195, 196
Codex Committee on Residues of
  Veterinary Drugs in Food, 108
Committee on Toxicity (UK), 63, 80,
  106
Cucurbitacins, 6, 17

Dietary protective factors, 16
Di-(2-ethylhexyl) adipate, 68, 79, 80,
  90
  hepatic effects, 80, 82
  in cheese, 69
Di-(2-ethylhexyl) phthalate, 79
  hepatic effects, 80, 81

Ergotism, 1

Flavour analysis, 192
Food Advisory Committee, 63
Food-chain contaminants, v
  natural toxicants, 1
  polychlorinated dibenzo-p-dioxins
    and polychlorinated
    dibenzofurans, 21
  polycyclic aromatic hydrocarbons, 61
Food contact materials, 65, 73
  analysis, 85
  examples, 73
  legislation, 74
  polymers, 68, 73, 79, 90
  toxicology, 73, 75
Food hazards, ranking, v, 2, 71
Food-production contaminants, v
  analysis, 85
  food contact materials, 65, 73, 85
  food taints, 191
  legislation, 64, 74, 100, 178
  pesticides, 64, 70, 177
  ranking, 71
  surveillance, 66
  veterinary drug residues, 65, 99, 109
Food Safety Act (1990), 79
Food Surveillance, UK Steering Group
  on Chemical Aspects, 62
  reports, 64, 79
  role, 62, 63
  working parties, 63
Food, tainted, 198

Glucosinolates, 7
  biological properties, 11, 12
  hydrolysis products, 11
  occurrence in brassicas, 10
  protective effect, 12
  UK daily intakes, 12
Glycoalkaloids, 8
  analysis, ELISA method, 9
  biological activity, 9
  occurrence in Solanaceae, 8
  potatoes, 8
  UK mean daily intakes, 10
Glycol softeners, analysis, 91
Glycyrrhizin, 6
Goat's milk, 15

Headspace analysis, 86, 88
  volatile food packaging
    contaminants, 87

Illudane sesquiterpenes, 14
Immunoaffinity chromatography, 116
Isoflavone glycosides, 8

Lectins, 18
Legumes, 5, 6, 13, 16
  and saponins, 13
Liquorice, 6

Maximum admissible concentrations
    (pesticides), 178
  EC Directive 80/778/EEC, 178
Maximum Residue Levels (MRLs)
    (veterinary products), 105, 109
  EC Directive 86/469/EEC, 106, 109
Medicines Act (1968), 99
  and veterinary drugs, 100
S-Methylcysteine sulphoxide, 7
Mineral hydrocarbons, analysis in
    food, 96
Ministry of Agriculture, Fisheries and
    Food, 100

Natural toxicants,
  control, 15
  intake, 4
  removal from diet, 16, 17
  research priorities, 5
  research strategy, 4
  risk, 2

Off-flavours, 191, 194

Peroxisome proliferators, 81
  biochemical markers, 82
  carcinogenic implications, 82
  dialkyl phthalates, 82
  hepatic effects, 81
  monoalkyl phthalates, 82
Peroxisomes, 81
Pesticides,
  acceptable daily intake, 67
  Advisory Committee on Pesticides,
    63
  analysis, 182
  daily intakes, 67
  EC regulations, 178
  GC–MS methods, 185
  immunoassay methods, 188
  in drinking water, 177
  maximum admissible concentrations
    (MACs), 178
  official analytical methods, 183
  residues in food, 64, 70
  SCA methods for pesticides, 184
  Standing Committee of Analysts
    (SCA), 183
  Water Supply Regulations 1989, 179
  WRc Aquacheck, 182
Phthalates, 79, 82, 90
  structure–activity relationships, 83
Plastic additives, 74, 76, 85
Plasticizers
  acetyltributyl citrate, 79, 90
  analysis, 90
  di-(2-ethylhexyl) adipate, 68, 79, 80,
    82, 90
  di-(2-ethylhexyl) phthalate, 79, 80,
    81
  epoxidized soya bean oil, 79, 94
  maximum daily intakes, 80
  migration, 80
  phthalates, 79, 82, 90
  polymeric, 79, 93
Plastic monomers
  analysis, 87, 92
  4-chlorophenyl sulphone, 92
  classification, 76
  EC directive 90/128/EEC, 79
Plastic oligomers
  PET, analysis, 95

Plastics, food contact, 65, 73, 79, 85
    BIBRA/British Plastics Federation
        Code of Practice, 74
    components, 73
    EC legislation, 75
    Food Surveillance Papers, 70, 79
    legislation, 74
    Scientific Committee for Food
        (SCF), 76
Polychlorinated biphenyls, 23
Polychlorinated dibenzo-*p*-dioxins and
        polychlorinated dibenzofurans, *see
        also* 2,3,7,8-TCDD
    analytical considerations, 39
    biogenic formation, 26
    chlorophenols, 23
    definition, 21
    environmental pathways, 30
    food chain, 32
    GC–MS analysis, 25
    incineration sources, 23
    intakes, health risk, 29
    isomer distribution, 22
    paper/paperboard, 37
    soil levels, 31
    sources, 23
    toxic equivalence schemes, 27
    toxicity, 26
    wood pulp bleaching, 24
Polychlorinated dibenzo-*p*-dioxins and
        polychlorinated dibenzofurans,
        determination in
    animals, 33
    eggs, 36
    fish, 36
    human adipose tissue, 38
    human milk, 38
    meat, 34
    non-human milk, 34
    vegetation, 32
Polycyclic aromatic hydrocarbons
    carcinogenic properties, 43
    daily intakes, 58
    heterocyclic analogues, 57
    isomer distribution, 42
    levels in food, 56, 58
    levels in smoked foods, 56
    metabolism, 44
    occurrence in food, 54, 58
    origin in food, 45

Polycyclic aromatic hydrocarbons,
        analytical techniques for food
    extraction, 46
    GC, 49
    HPLC, 49
    SFC, 52
    spectroscopic identification, 52
    TLC, 48
Polymeric plasticizers, analysis, 93
Potato, 8
    damaged, 6
    glycoalkaloids, 8
    Lenapé cultivar, 7
    Solanum chacoense, 7
    Solanum vernii, 7
Psoralens, 6
Ptaquiloside, 14
    in bracken, 14
Pyrrolizidine alkaloids, 6

Quinoa, 17

Rapeseed, 7, 18
Regenerated cellulose film, 82, 91

Saponins, 7
    biological properties, 13, 14
    daily intakes, 14
    occurrence in legumes, 13
    soya, 13
Scientific Committee for Food (SCF),
        76
    and plastic ingredients, 76, 78
        toxicity tests, timetable, 78
Seveso, 22, 27, 32
Size-exclusion chromatography, 90
Solanaceae, 8
    and glycoalkaloids, 8
α-Solanine, 8, 13
Solid-phase extraction, 115, 185
Soya, 8, 14
Soyasaponins, 13
Squash, 6
Standing Committee of Analysts, 183
Steering Group on Chemical Aspects
        of Food Surveillance, 62
Stilbene residues, 67, 135
Structure–activity relationships, 80,
        83, 84

Tainted food, 198
  examples, 198, 199
  yeast, 193
Tainted food, contaminants
  chloroanisoles, 195, 196, 198
  chlorophenols, 195, 196
  N,N-dimethylaniline, 196
  gas chromatographic analysis, 192
  geosmin, 195
  indole, 193
  mass spectrometric analysis, 193
  methyl vinyl ketone, 197
  2-methoxy-3-isopropylpyrazine, 195
  2-methylborneol, 195
  naphthalene, 197
  styrene, 197
2,3,7,8-TCDD
  carcinogenic risk, 26
  2,4,5-trichlorophenoxyacetic acid,
    22, 23
Thermosets and thermoplastics, 73

Vegetarians, 13, 14
Veterinary drugs
  acepromazine, 167
  β-agonists, 127
  aklomide, 149
  albendazole, 126
  aminoglycosides, 123
  ampicillin, 131
  amprolium, 169
  anabolic steroids, 135
  androgens, 137
  azaperol, 167
  azaperone, 167
  benzimidazoles, 125
  benzyl penicillin, 130, 131
  cambendazole, 126
  carazolol, 167
  carbadox, 132
  cephapirin, 131
  chloramphenicol, 133
  chlorpromazine, 167
  cimaterol, 128
  clenbuterol, 128
  clopidol, 171
  cloxacillin, 130
  dapsone, 172
  destomycin A, 125
  detomidine, 168

  dexamethasone, 138
  diaminopyrimidines, 157
  diaveridine, 162
  diethylstilbestrol, 135
  dihydrostreptomycin, 124
  dimetridazole, 152
  dinitolmide, 149
  dinsed, 152
  dyes, 172
  enrofloxacin, 157
  erythromycin, 146
  ethopabate, 173
  febantel, 126
  fenbendazole, 125
  flumequine, 154
  furaltadone, 149
  furazolidone, 148, 150
  gentamicin, 123
  gestagens, 136
  halofuginone, 173
  hexestrol, 139
  hormones, 135
  5-hydroxythiabendazole, 125
  hygromycin B, 125
  ionophore polyethers, 141
  ipronidazole, 153
  ivermectin, 144
  β-lactams, 129
  lasalocid, 142
  Leucogentian Violet, 172
  levamisole, 173
  lincomycin, 124
  macrocyclic lactones, 144
  macrolide antibiotics, 145
  Malachite Green, 172
  mebendazole, 126
  melengestrol acetate, 137
  methylthiouracil, 167
  metronidazole, 153
  monensin, 142
  morantel, 174
  nalidixic acid, 154
  narasin, 143
  neomycin, 124
  nicarbazin, 151
  nitrofurans, 147
  nitrofurantoin, 149
  nitrofurazone, 149
  nitroimidazoles, 152
  nitromide, 149

nitroxynil, 151
novobiocin, 175
oestrogens, 137, 139
olaquindox, 132
oleandomycin, 146
ormetoprim, 162
oxfendazole, 125
oxolinic acid, 154
penicillins, 129
phenoxypenicillin, 131
piromidic acid, 155
propiopromazine, 168
pyrantel, 174
pyrimethamine, 162
quinolone carboxylic acids, 154
quinoxaline-2-carboxylic acid, 132
rafoxanide, 175
ronidazole, 153
salbutamol, 128, 129
salinomycin, 143
sedecamycin, 147
spiramycin, 146
stilbenes, 135
streptomycin, 125
sulphadiazine, 158
sulphadimethoxine, 158
sulphamethazine, 157
sulphanitran, 149
sulphaquinoxaline, 159
sulphathiazole, 157
sulphonamides, 157
terbutaline, 128, 129
tetracyclines, 163–166
thiabendazole, 125
thiamphenicol, 134
thiouracil, 167
thyrostats, 167
tranquilizers and sedatives, 167
trenbolone, 137
triclabendazole, 127
trimethoprim, 161
tylosin, 146

virginiamycin, 175
xylazine, 167
zeranol, 137
Veterinary drug residues,
    determination
    immunochemical methods, 116
    limits of detection (LoDs), 113, 121
    Matrix Solid Phase Dispersion
      (MSPD), 115
    microbiological methods, 119
    multi-residue methods, 122
    physico-chemical methods, 114
    solid-phase extraction (SPE), 115
Veterinary Medicines Directorate,
    100, 101
Veterinary products
    banned substances, 106, 127, 135
    EC legislation, 103, 106, 109
    feeding-stuffs, 103
    Maximum Residue Levels (MRLs),
      105, 109–113
    monitoring, 106
    product licence, 100
    residues, 104, 108
    Veterinary Medicines Directorate,
      100, 101
    Veterinary Products Committee, 63,
      100, 110

Water
    analysis for pesticides, 182
    EC Drinking Water Directive, 178
    Guidance Notes for analysis, 180
    tainted food, 195
Water Supply Regulations, 179
    analytical methods, performance
      requirements, 180
    analytical quality control, 182
    sampling, 179
Wood pulp bleaching, 24

Zucchini, 6